The Science of Human Milk

The Science of Human Milk

JOSEF NEU
Department of Pediatrics, Division of Neonatology,
University of Florida, Gainesville, FL, United States

ELSEVIER
ACADEMIC PRESS
An imprint of Elsevier

Academic Press is an imprint of Elsevier
125 London Wall, London EC2Y 5AS, United Kingdom
525 B Street, Suite 1650, San Diego, CA 92101, United States
50 Hampshire Street, 5th Floor, Cambridge, MA 02139, United States
The Boulevard, Langford Lane, Kidlington, Oxford OX5 1GB, United Kingdom

ISBN: 978-0-323-99913-7

For Information on all Academic Press publications
visit our website at https://www.elsevier.com/books-and-journals

Publisher: Nikki P. Levy
Acquisitions Editor: Nicole Denis
Editorial Project Manager: Catherine Costello
Production Project Manager: Bharatwaj Varatharajan
Cover Designer: Miles Hitchen

Typeset by MPS Limited, Chennai, India

Contents

Preface

Humans, as mammals, have evolved over the millennia to utilize their mother's milk as a primary food after parturition. For various reasons, including sociocultural and economic, breastfeeding for many has been replaced by other means of nourishment for infants. Numerous studies have demonstrated the short- and long-term benefits of breastfeeding for both mother and infant.

Concurrently, emphasis has been placed on exclusive breastfeeding for prolonged periods (sometimes years after birth), the use of pooled banked breast milk from several donors for not only preterm infants but also term infants whose mothers may not be able to provide their own milk, supplementation with various agents, the use of pumped and frozen breast milk, and adding certain components to formulas that are present in breast milk. Not all of these are based on sound science.

As stated by the esteemed physicist, Richard Feynman, "Religion is a culture of faith; Science is a culture of doubt." Despite many aspects of breastfeeding and providing human milk being based on a solid scientific basis, several practices may not be as solid and require scrutiny to provide benefit and also prevent harm if not used correctly. Overstatements suggesting that nonbreastfed infants will become stupid, obese, hate their parents, and spend considerable time in hospitals may actually be detrimental. Such statements are not based on sound science.

This book will address some of these issues and try to present a balanced approach that relies on studies of the composition of human milk, the risks, and benefits to certain infants. It will address many of the aspects related to composition. It will address several aspects of human milk feeding for special situations such as infants born preterm and cared for in neonatal intensive care units. It also addresses some of the aspects of mothers breastfeeding their own infants that are unique and will be very difficult to replicate even with the use of highly sophisticated emerging technologies.

Josef Neu

CHAPTER ONE

Evolution of human milk and breastfeeding

The love for all living creatures is the most noble attribute of man

Charles Darwin

Key points

1. There are special factors that are used to describe a mammal.
2. Intermediate taxa such as the monotremes and marsupials provide evolutionary links between egg-laying animals and those that give birth to their offspring and provide milk as a nutrient.
3. Numerous differences, as well as similarities, exist in the composition of milk between species of mammals.

Introduction

Why include a chapter about evolutionary aspects of breastfeeding and human milk in this book? Obtaining a better understanding of the past helps us appreciate how we as human mammals emerged into a successful species with relatively high cognitive and physical abilities. One of the major aspects of this relates to how we feed our offspring after birth with a fluid that is highly adapted for nutritional and protective functions.

Nutrition is critical for the survival of all living organisms. Even single-celled animals have needs to assimilate nutrients, which they accomplish through their cell membranes. As multicellular organisms evolved, different cell types differentiated and specialized to not only provide nutrition

The Science of Human Milk.
DOI: https://doi.org/10.1016/B978-0-323-99913-7.00006-X

for themselves but also to catalyze nutrition for other components of the organism. As these multicellular organisms evolved into what we know as the animal kingdom, reproduction produced offspring that had nutritional requirements. Some required help from the parent, whereas others did not, and nutritional independence in offspring was highly variable. One such primitive animal is the hydra, a simple, multicellular animal with several tentacles around its mouth used for the ingestion of food. The tentacles entangle small aquatic animals and kill them with their stinging cells and push them into the mouth. Inside the body cavity the digestive juices secreted by the surrounding cells help to digest the food. This is then absorbed through the cavity walls and assimilated in the cells.

When food is plentiful, many *Hydras* reproduce asexually by budding. The buds form from the body wall, grow into miniature adults and break away when mature. A new bud can form every 2 days. Since they are already mature adults after budding, no special process needs to occur that provides additional nutrition from the parent after the budding process. When conditions are harsh, often before winter or in poor feeding conditions, sexual reproduction occurs in some *Hydra*. Swellings in the body wall develop into either ovaries or testes. The testes release free-swimming gametes into the water, and these can fertilize the egg in the ovary of another individual. The fertilized eggs secrete a tough outer coating, and, as the adult dies (due to starvation or cold), these resting eggs fall to the bottom of the lake or pond to await better conditions, whereupon they hatch into nymph *Hydra* but do not require a parent for nutritional support. As seen in this example, providing food to offspring by a parent can be accomplished in several ways.

Millions of years ago, another of the major means for primitive animals was to form an egg from a zygote, allow it to grow in the parent for a time, then expel it into the external environment There it would continue to require care for survival. The contents of the egg provided transitional food for the growing offspring. Most birds and reptiles continue to transition their offspring to their adult diets in this manner.

In modern times, milk provides a unique transitional food for the offspring during adaptation to the external environment for mammals, who are relative newcomers to the evolutionary tree of life (see Fig. 1.1). More on mammals later, but first, how did milk evolve and why did it evolve in the way it did to the current highly nutritious fluid upon which most mammals appear to be dependent in early life.

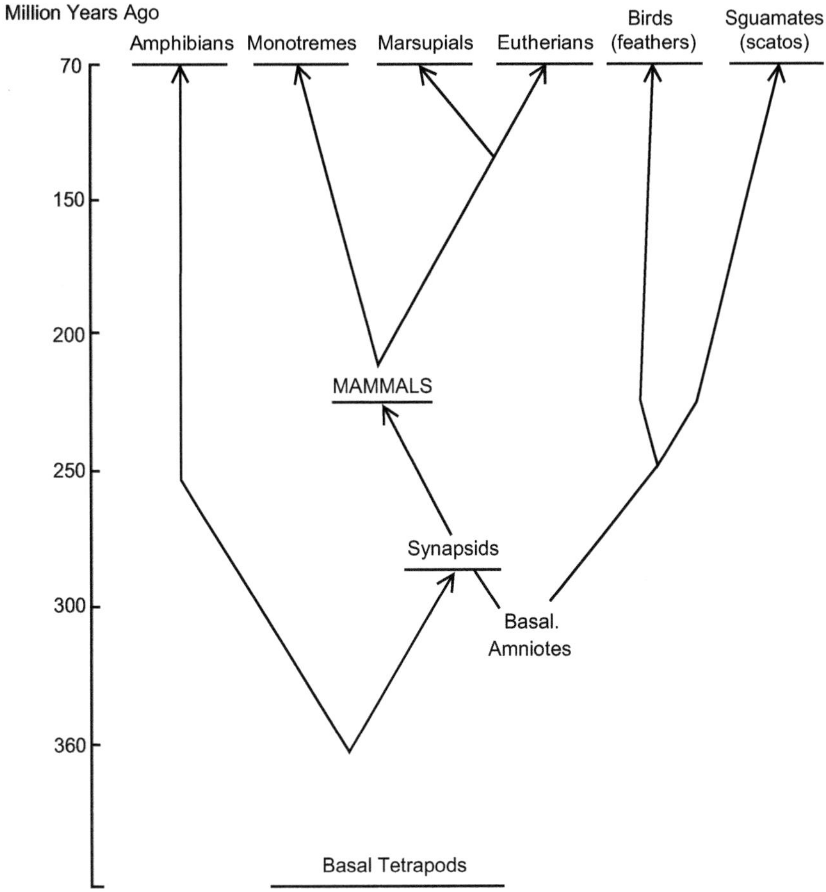

Figure 1.1 Evolutionary perspective. *Adapted from Lefevre C, Sharp J, Nicholas K. Evolution of lactation: ancient origin and extreme adaptations of the lactation system. Annu Rev Genomics Hum Genet 2010;11:219–38.*

Origins of milk

The evolution of milk has ancient origins [1,2]. Fig. 1.1 illustrates several of the major evolutionary aspects of milk development.

Looking from the bottom to the top, approximately 400 million years ago, sea creatures ("basal tetrapods"—animals with 4 legs) that laid eggs, which were protected by the moist environment of the sea, developed the means of moving onto land. The types of eggs that were laid began to evolve, and the land animals needed to be able to protect these eggs.

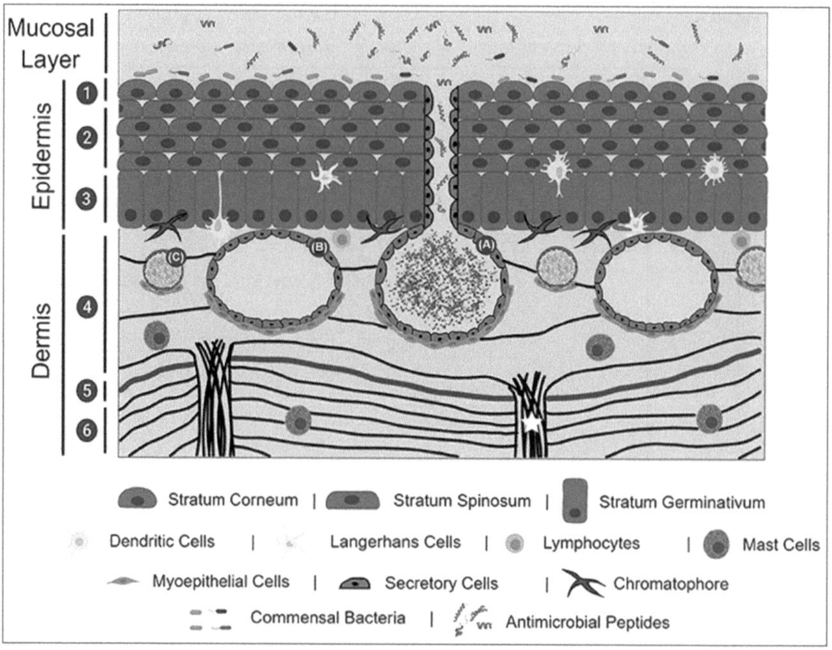

Figure 1.2 Amphibian epidermal layer with secretions resembling some aspects of mammalian milk. Glands within the dermal layer of frog skin include granular glands (A), mucosal glands (B), and small mixed glands (C). These secrete several compounds, including mucus and antimicrobial peptides. Commensal bacteria overlay the frog skin layers, forming the microbiological barrier. *From https://www.frontiersin.org/articles/10.3389/fimmu.2018.03128/full.*

A branch evolved into a group that could live both on dry land and water ("amphi"-bians), with primitive lungs. They laid eggs without shells and the larval stage of these amphibians still have gills like their more primitive ancestors. Amphibians retain their ability to lay eggs, but many species also have secretions from their skin that have very potent immunologic properties (Fig. 1.2) [3] like that of mammalian milk today.

Some of the land tetrapod species evolved into amniotes, most of which are completely terrestrial tetrapods. Amniotes are a clade of tetrapod vertebrates that comprise sauropsids (including all reptiles and birds, and extinct parareptiles and nonavian dinosaurs) and synapsids (including pelycosaurs and therapsids such as mammals).

The key innovation in amniotes over amphibians is the amnion, which enables the eggs to retain their aqueous contents on land, rather than needing to stay in water. The amnion is a watertight membrane that immediately surrounds the developing embryo. This allowed tetrapods to lay watertight amniotic eggs on land. The amnion is a gelatinous capsule that contains the developing embryo that is permeable to water and gases. It is surmised that this protection was accomplished via close approximation of these eggs to the skins of these animals, from which liquids were secreted that kept the eggs moist and provided protective chemicals with immunologic properties. The eggs in many of these primitive animals did not include a calcified eggshell such as that in modern-day lizards and birds. Rather it was parchment-like. It is thought that the skin secretions were rich in antimicrobial and other protective compounds and thus were a primitive form of milk.

The other branch that evolved via the tetrapod and sauropsid stages evolved over approximately 100 million years ago into mammals (Fig. 1.1). The early mammals, the synapsids, developed glands in their skins that replaced some of the scale-like structures present when not yet evolved into land animals. These sweat glands are sometimes associated with hairy structures to form what is a pilosebaceous unit. Presumably, these secretions met the parchment-like eggshells and offered protection.

With further evolution, animals that still did not have placentas and laid eggs early in gestation evolved into the monotremes (Fig. 1.3, below). The earliest split in mammalian phylogeny occurred about 200 million years ago when the Proterians (monotremes) separated from the marsupials (Figs. 1.4 and 1.5). Current-day examples of the monotremes include the platypus and echidans, found primarily on the Australasian continent. These animals utilize apocrine glands on the skin that are organized into small mammary patches or areola consisting of about 100–200 sebaceous

Figure 1.3 Monotremes.

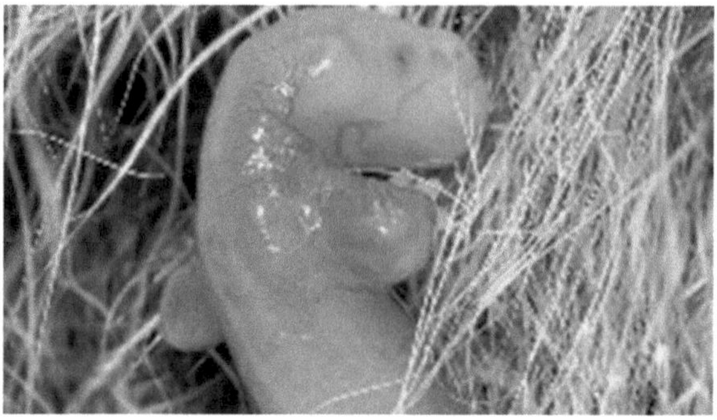

Figure 1.4 Marsupial newborn.

Figure 1.5 Continuation of marsupial infant development.

units without a nipple from which the hatched monotreme infant receives nutrition. The young monotreme after hatching, stays near the mother, where secretions may be suckled from the hairs, thereby providing nutrition for the monotreme infants.

This sebaceous unit secretion mechanism evolved so that in the modern age similar secretions are no longer a characteristic feature, and lactation has become the near-universal mode of nourishing the young in more modern mammalian species.

What is a mammal?

Carl Linnaeus, the Swedish botanist, and physician who is well known for his classification (taxonomy) of living organisms defined a mammal as "an animal that suckles its young," 1826,—from Latin mamma "breast." Details—Caroli Linnaei. . .Systema naturae per regna tria naturae: secundum classes, ordines, genera, species, cum characteribus, differentiis, synonymis, locis—Biodiversity Heritage Library (biodiversitylibrary.org). Except for the few egg-laying species described above, all bear live young and have the mammary gland for the young to suck. Other features of mammals include:

- hair,
- lower jaw that is hinged directly to the skull,
- middle ear bones,
- diaphragm that separates heart and lungs from abdominal cavity,
- left aortic arch,
- red blood cells without nuclei,
- warm-blooded and breathe air.

Next, we will provide a further description of the main categories of mammals including the monotremes, marsupials, and the eutherians, through which the process of modern-day lactation evolved.

Let us start with what we mean by lactation. Lactation describes the secretion of a nutritious fluid called milk from the mammary glands. Lactation naturally occurs with all sexually mature female mammals. The process of feeding milk in all animals (including humans) is called *nursing*, and in humans it is also called *breastfeeding*. Nearly every species of mammal has nipples; except for monotremes, egg-laying mammals, which instead release milk through ducts in the abdomen.

The chief function of lactation is to provide nutrition and immune protection to the young after birth. Due to lactation, the mother–young pair can survive even if food is scarce or too hard for the young to attain, expanding the environmental conditions the species can withstand. In almost all mammals, lactation induces a period of infertility (in humans), which serves to provide the optimal birth spacing for survival of the offspring.

The oldest taxa of the class Mammalia are the monotremes (Fig. 1.3). The word monotreme literally means "one hole," referring to the single duct (the cloaca) for their urinary, defecatory, and reproductive systems. Like reptiles, monotremes have a single cloaca.

Monotremes are the only egg-laying mammals

The egg is only the size of a grape, and the shell is very thin, fragile, and has a leathery texture. A type of monotreme, Echidnas, only breeds during July–September every year, so that is the best time to spot them in the wild, as they are most active.

As mentioned, these animals lay eggs but have no placenta or nipples. Today, monotremes are found in the wild only in Australia and New Guinea. There are only five living monotreme species: the duck-billed platypus and four species of echidna (also known as spiny anteaters). Monotremes do not exhibit much diversity; they are very primitive compared to other mammals because like reptiles they lay eggs. In several other respects, monotremes are different than other mammals. They have modified snouts or beaks, and adults do not have teeth. Like other mammals, however, monotremes have three middle ear bones, a single bone in their lower jaw, hair, high metabolic rates, and produce milk to nourish the young.

With additional advancement, another form of mammal evolved, the marsupial. All marsupials alive today are endemic to Australasia, Indonesia, and the Americas. A distinctive characteristic common to marsupials is that the young are carried in a pouch. Living marsupials include opossums, Tasmanian devils, kangaroos, koalas, wombats, wallabies, and bandicoots.

Marsupials give birth to relatively undeveloped young that often reside in a pouch located on their mothers' abdomen for a certain amount of time. The word *marsupial* comes from *marsupium*, the technical term for the abdominal pouch.

In marsupials, there is also a form of mammary gland that is derived from hair follicles and sebaceous glands, but a nipple is usually present. It is notable that in marsupials, the placenta, as in monotremes, is not very well formed, and in marsupials the pouch along with the mammary glands takes over the role of the placenta. Marsupials are characterized by what some may consider a premature birth when compared to other mammals. The newborns continue to develop while attached to the nipples on the mother's lower belly (Fig. 1.6).

The pouch—or marsupium, from which the group takes its name—is a flap of skin covering the nipples. This can be seen in the kangaroo mother and her joey at the left. Wombats and koalas are other examples of marsupial animals. The marsupium is prominent in many species, but it is not a universal feature. In some species the nipples are fully exposed. They may be bounded by remnants of a pouch. The young remain firmly attached to the teats for a period that approximates the latter part of the development of the fetus in the womb of a placental mammal (eutherian—Fig. 1.6).

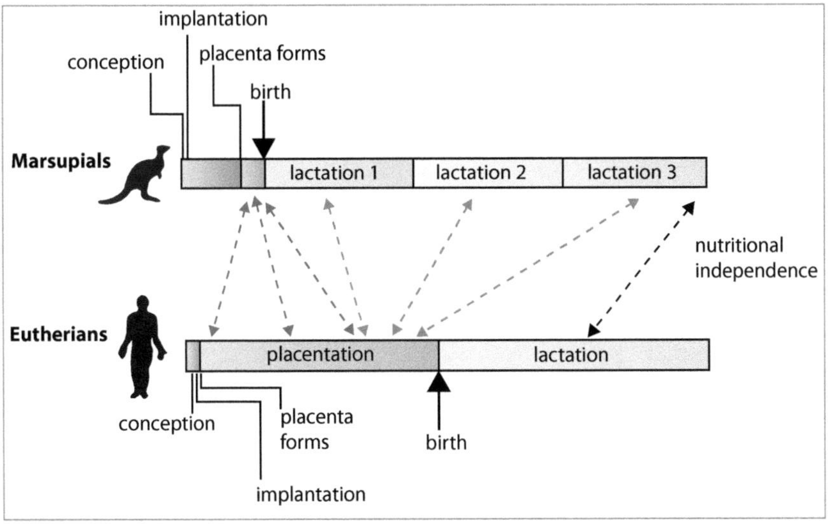

Figure 1.6 Eutherians—placental mammals.

Eutherians

Eutherians comprise most of the taxa of mammals that live on Earth today. In eutherian mammals (of which humans are a representative), a well-formed placenta is present and there is well-formed nipple development. Lactation provides essential nutrients required by the mammalian neonate (Fig. 1.7).

In eutherians, the energy invested by the mother in rearing young during placentation versus placentation is nearly equal. In the brief gestation in marsupials, the placenta forms late in pregnancy, but lactation is extended. Some of the genes that underlie placental functions in eutherians (red arrows) are expressed during lactation in marsupials (blue arrows), including various conserved components of lactation itself (black arrow).

In modern mammals, the composition of milk may vary tremendously from one species to the next, as does the quantity. Some modern mammals receive very small quantities of milk shortly after birth and this quantity can range to over 200 kg of milk per day in whales. Some others will store nutrients during one period, store them in their tissues and then restore this to milk and provide them to their young later. Some bears will not eat for 2 months even though they lactate and feed their young during this time.

There are numerous fascinating facts about milk from these early taxa of mammals, including proteins such as caseins that are highly structurally

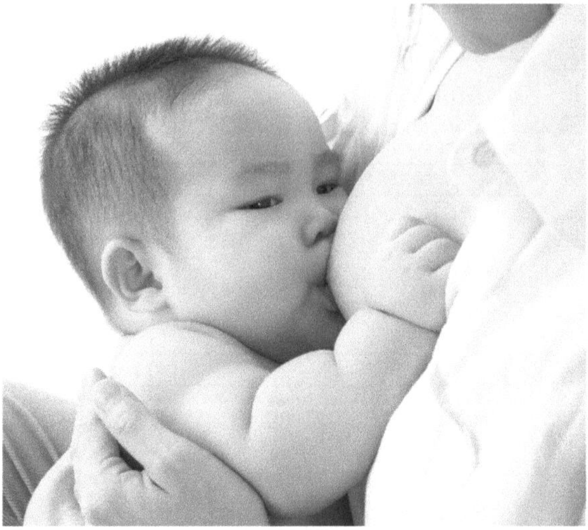

Figure 1.7 Different reproductive strategies of eutherian versus marsupial mammals have been compared. *From Lefevre C, Sharp J, Nicholas K. Evolution of lactation: ancient origin and extreme adaptations of the lactation system. Annu Rev Genomics Hum Genet 2010;11:219–38.*

conserved between monotremes, marsupials, and eutherians. Many of these have major immunologic functions. In fact, recent studies are closely scrutinizing a protein found in platypus "milk" with powerful immunologic potency for use when microbes develop resistance to antibiotics [4].

Evolution of lactation

Lactation is the process of nourishing the young mammal with milk secretion via the mammary gland [5]. As seen in the previous section, lactation has an ancient origin. Analyses of milk proteins and other components show that many of these are highly conserved over these species. Furthermore, there can be extreme adaptations in the lactation system from one type of mammal compared to others. One example is the complete lack of lactose synthesis in fur seals compared to other mammals [6].

Mammary gland origin

The structure of the mammary gland is like that of skin glands and hair follicles. The exact pathways of mammary gland evolution remain

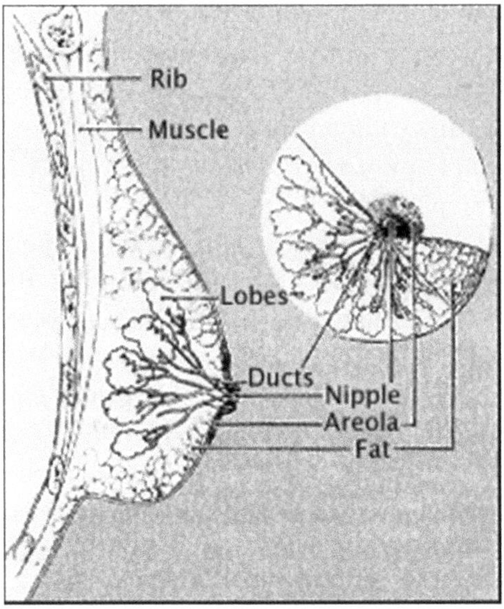

Figure 1.8 Breast anatomy. *Image courtesy NCI/NIH.*

somewhat speculative, but as previously mentioned, a derivation from ancestral skin glands of the apocrine type that coevolved with the development of fur appears to represent a reasonable hypothesis. Approximately 200 million years ago, there occurred an evolution of a complex lactation system with a highly differentiated mammary gland that specialized in the production of complex secretions including some that were specific for milk and were also protective for the offspring (Fig. 1.8).

Anatomy, development, and physiology of the breast

The breast is a mass of glandular, fibrous, and fatty tissues positioned over the pectoral muscles of the chest wall and attached to the chest wall by fibrous strands called Cooper's ligaments. The breast has a soft consistency because of the fatty tissue that extends through the breast and surrounding tissue. The glandular tissue of the breast is comprised of the milk-producing glands at the ends of the lobes called lobules. There are also milk passages called ducts. The ducts widen toward the nipple, where they form a sac called the ampulla. During the process of lactation, milk is produced from the ends of the lobules, where in turn, it is transferred to the nipple and ultimately the infant.

Female breasts do not begin growing until puberty, which usually begins for women around age 10 or 11. At that time, breasts begin to respond to hormonal changes in the body. Progesterone and estrogen are the main hormones signaling the development of the glandular breast tissue.

During pregnancy, numerous breast changes occur. They increase in size quickly. The nipples become sore a couple of weeks after conception, while the breasts become tender. During this time the nipples also become larger and more erect. Estrogen stimulates the growth of the ducts and progesterone cause the glandular tissue to expand, while the blood vessels within the breast enlarge.

Prolactin and oxytocin are two of the most important hormones involved in lactation [7]. The body begins producing prolactin approximately 8 weeks after conception and is related to behavioral changes in the mother that includes a tranquilizing effect. Prolactin is sometimes referred to as the "mothering hormone." As the pregnancy progresses, the levels of prolactin steadily increase, peaking when the woman gives birth. As the body produces more and more prolactin, high levels of estrogen and progesterone block some of the prolactin receptors and inhibit milk production until after the baby is born.

After birth, estrogen, progesterone, and prolactin levels decline. The breasts will usually begin to produce milk 3–5 days after a woman has given birth. Prior to the production of milk, for the first 2–3 days after birth, colostrum is produced. This contains high levels of antibodies to help protect the infant against infections.

The other hormone responsible for milk production, oxytocin, aids in bringing milk out of the nipples. The breasts will continue to produce milk until the mother stops breastfeeding and/or expresses milk from her breasts. Several months may be required for milk production to completely stop.

Next, we will go into more detail about the physiology as well as the different stages of lactation

Lactation physiology

Lactogenesis involves the maturation of alveolar cells in the breast so that they are able to secrete milk [8]. It involves two stages which include secretory initiation and activation.

- Secretory initiation occurs during the second half of pregnancy. High levels of progesterone are produced in the placenta and by week 16 small amounts of milk can be secreted.

- Secretory activation occurs after delivery when large amounts of milk begin to be secreted. Progesterone levels drop precipitously with the removal of the placenta at delivery, but high levels of prolactin persist to cause stimulation. By the second to third day after delivery, swelling of the breast occurs along with the production of large quantities of milk. If this is the first pregnancy, often the secretory activity stage is slightly delayed, and the amount of milk produced is less than in women who have previously given birth. Lower milk volume is also observed in women who had cesarean births compared with those who delivered vaginally.

The removal of milk stimulates further lactation. Stimulation of the nipple triggers prolactin release by the anterior pituitary and oxytocin from the posterior pituitary glands. Prolactin stimulates mammary gland ductal growth. It induces epithelial cells to proliferate and stimulates milk protein synthesis. Emptying of the breast by the infant's suckling stimulates nerve endings, which in turn, enhances milk secretion.

In addition to milk production, ejection also needs to occur, and this is aided by the action of oxytocin, which is involved in the letdown reflex. Stimulation of the nipple–areolar complex by suckling signals to the hypothalamus triggering the release of oxytocin. This results in the contraction of muscle cells called myoepithelial cells, which force milk into the ducts from the alveolar lumens and out through the nipple. Oxytocin also has a psychological effect, which includes inducing a state of calm and reducing stress.

If milk is not removed, elevated pressure in the mammary gland causes stimulation of inhibitors of lactation, which reduce milk production and initiate mammary involution. This feedback inhibitor of lactation regulates how much milk is produced which is determined by how much the baby takes.

Diversity of lactation strategies

Eutherian mammals have extended intrauterine development of the young and except for initial colostrum, produce a milk of relatively constant composition. Marsupials, on the other hand, after a short pregnancy, give birth to young that are dependent upon a constantly changing milk composition for development during an extended lactation period. Some even produce different composition of milk from adjacent mammary glands, a phenomenon called asynchronous concurrent lactation [9].

Lactation can vary dramatically between mammalian species [10]. For example, lactation length in some earless seals may be only 3–5 days to several years in some primates. These differences may be related to differential constraints related to the energetic, nutritional or socio-protective drivers of lactation in different lineages.

Diversity of milk composition

The macronutrients (carbohydrates, fats, and proteins) are considerably different between species [11]. Certain seals and walruses have a 50% milk fat content while rhinoceros' milk has 0% fat. Some of these differences may relate to maternal diet. The composition of milk in marsupials drastically changes during lactation, whereas in humans, the milk produced during the first 5 days (colostrum) differs significantly from the relatively constant transitional or mature milk (to be discussed in later chapters in more detail).

Can there be lactation without pregnancy?

In humans, lactation and relactation can be induced. This can be helpful when there are adoptive mothers or wet nurses. It appears plausible that the possibility of lactation in women (or females of nonhuman species) who are not biological mothers may confer evolutionary advantages, especially in groups with high maternal mortality and tight social bonds.

Lactation can be induced in humans by a combination of physical and psychological stimulation, by drugs, or by a combination of those methods. These involve the use of contraceptive pills to mimic the hormone levels of pregnancy with domperidone, a dopamine antagonist, which also raises the level of prolactin in the human body. After the use of the domperidone and contraceptives, an electric breast pump is used to induce and continue milk production, until the infant successfully breastfeeds.

Conclusion

While detailed discussions of the various compositions of milk from different mammalian species are beyond the scope of this book, it is important to note that despite numerous differences due to environmental adaptations and evolutionary pressures different species have faced, there is also evolutionary conservation of various proteins, lipids, carbohydrates, and immunoprotective agents between these species.

References

[1] Lefevre C, Sharp J, Nicholas K. Evolution of lactation: ancient origin and extreme adaptations of the lactation system. Annu Rev Genomics Hum Genet 2010;11: 219–38.
[2] Oftedal OT. The evolution of lactation in mammalian species. Nestle Nutr Inst Workshop Ser 2020;94:1–10.
[3] Varga JFA, Bui-Marinos MP, Katzenback BA. Frog skin innate immune defences: sensing and surviving pathogens. Front Immunol 2019;9:3128.
[4] Newman J, Sharp JA, Enjapoori AK, Bentley J, Nicholas KR, Adams TE, et al. Structural characterization of a novel monotreme-specific protein with antimicrobial activity from the milk of the platypus. Acta Crystallogr F Struct Biol Commun 2018;74(Pt 1):39–45.
[5] Brines J, Billeaud C. Breast-feeding from an evolutionary perspective. Healthcare (Basel) 2021;9(11):1458.
[6] Sharp JA, Lefèvre C, Nicholas KR. Lack of functional alpha-lactalbumin prevents involution in Cape fur seals and identifies the protein as an apoptotic milk factor in mammary gland involution. BMC Biol 2008;6(1):1–15.
[7] Uvnäs-Moberg K, Widström AM, Werner S, Matthiesen AS, Winberg J. Oxytocin and prolactin levels in breast-feeding women. Oxytocin and prolactin levels in breast-feeding women. Acta Obstet Gynecol Scand 1990;69(4):301–6.
[8] Pillay J, Davis TJ. Physiology, lactation. Treasure Island (FL): StatPearls Publishing; 2022.
[9] Nicholas K, Simpson K, Wilson M, Trott J, Shaw D. The tammar wallaby: a model to study putative autocrine-induced changes in milk composition. J Mammary Gland Biol Neoplasia 1997;2(3):299–310.
[10] McClellan HL, Miller SJ, Hartmann PE. Evolution of lactation: nutrition v. protection with special reference to five mammalian species. Nutr Res Rev 2008;21(2): 97–116.
[11] Pietrzak-Fiećko R, Kamelska-Sadowska AM. The comparison of nutritional value of human milk with other mammals' milk. Nutrients. 2020;12(5):1404.

CHAPTER TWO

Macronutrients and micronutrients

"One should eat to live; not live to eat"

Socrates

"The food we eat goes beyond its macronutrients of carbohydrates, fat, and protein. It's information. It interacts with and instructs our genome with every mouthful, changing genetic expression"

David Perlmutter

Key points

1. Carbohydrates, lipids, and proteins constitute the macronutrients of milk. These provide the major energy, metabolic, and growth needs of the infant.
2. The composition of human milk is dynamic. It reflects diurnal patterns, lactational period (colostral, transitional, and mature), and changes during the time of each individual feeding.
3. Maternal diet can play a major role in both macronutrient and micronutrient composition of milk.

Introduction

Human milk contains carbohydrates, proteins, and lipids that are termed macronutrients and provides nutrition that meets the energy, metabolic, and growth needs for infants. There is another type of carbohydrate in human milk termed human milk oligosaccharides, which are present in high quantities, but not absorbed or directly used for energy or metabolic growth needs. However, these are thought to have important

The Science of Human Milk.
DOI: https://doi.org/10.1016/B978-0-323-99913-7.00002-2

properties and are discussed in more detail later. Human milk also contains micronutrients, including vitamins and minerals required for optimization of metabolism, growth, and physiologic homeostasis. Here we will focus not only on macronutrients but also provide a brief description of the vitamins and minerals found in human milk. The macronutrients' relative contributions to total energy intake are approximately 44.5% carbohydrate, 43.9% fat, and 8.4% protein [1].

Carbohydrates

Mature human milk in terms of mass contains approximately 7% carbohydrate, with the majority being the disaccharide lactose. Lactose is hydrolyzed into its two component monosaccharides glucose and galactose at the brush border of the intestinal epithelial cells, where they are subsequently absorbed [2–4]. Other carbohydrates also play a role in infant health. The oligosaccharides (discussed in greater detail in Chapter Six: Human Milk Oligosaccharides) are neither digested nor absorbed by the human host but are hydrolyzed by microbes and release a rich source of bioactive metabolites [5]. These exhibit properties similar to fiber found in other foods that are also thought to play a significant role in infant health but are consumed by older infants, children, or adults.

Factors controlling carbohydrate composition of milk

Several factors determine the composition of breast milk carbohydrates. One of the major factors is maternal diet. The high prevalence of added sugars in modern diets and the increased intake during pregnancy has focused attention on these. One of the monosaccharides found in human milk is fructose, which in recent years has increased in human milk composition because it has become ubiquitous in the food supply and is often ingested by the mothers [6,7]. The physiologic and health consequences of fructose in human milk remain poorly understood, but it is present in appreciable quantities in the form of maternally ingested high-fructose corn syrup [6]. Studies have shown that a beverage containing these had no effect on lactose or glucose concentrations, but significantly increased fructose concentrations in breast milk. This raises several questions about the effect of these on the early development and health of the infant.

Theoretical concerns have been raised about whether maternal intake during pregnancy may result in obesity in her offspring later in life [6,7]. In a small proof-of-concept study, high fructose was significantly associated with higher body weight, and this effect was distributed across all components of body composition (i.e., fat mass, lean mass, and bone mass) [7]. Although these results are preliminary, they raise the possibility that increased consumption of corn syrup solids (found in many sweetened foods) may have a significant adverse metabolic effect on the breastfed infant. Furthermore, another recent study showed neurocognitive deficits at 2 years of age in infants whose mothers consumed high levels of sugar-sweetened beverages during lactation [8].

There is not much data relating the role of carbohydrate composition in breast milk to somatic growth of the infant. Some early studies that evaluated lactose and glucose concentrations in relation to infant weight and body composition [9,10] showed that higher total carbohydrate concentrations were associated with greater fat mass and fat mass to fat-free mass ratios. Higher lactose calculated daily intakes were associated with greater fat mass and fat mass to fat-free mass ratios. Concentrations and intakes of human milk carbohydrates differentially influence the development of infant body composition in the first 12 months postpartum and may potentially influence the risk of later obesity [10].

Carbohydrates in breast milk also affect neurodevelopmental outcomes. Of the individual constituents in breast milk that support brain development, carbohydrates are of major importance, particularly when considering oligosaccharides. These will be discussed in greater detail in a later chapter. In animals, oligosaccharides have been shown to influence brain function through several mechanisms and include enhanced learning and memory [11]; sialylated and fucosylated oligosaccharides act as prebiotics in the gut microbiome to support fermentation products for brain signaling. An increased quantity of these oligosaccharide-stimulated microbes appears to enhance cognitive performance [12].

Proteins

After carbohydrates and lipids, proteins are the third most abundant macronutrient in breast milk. Proteins provide amino acids indispensable as building blocks for growth but also bioactive proteins and peptides essential for many functions. Specialized bioactive proteins will be discussed in detail in Chapter Three, Bioactive Components of Human Milk.

Casein and whey composition

The estimated number of human milk proteins is approximately 400. These are comprised of three major categories: caseins, whey, and mucins [13].

Caseins are phosphoproteins that are commonly found in mammalian milk, comprising about 80% of the proteins in cow's milk and between 20% and 60% of the proteins in human milk. Sheep and Cow milk have higher casein content than other types of milk with human milk having particularly low casein content. Casein contains a high number of proline amino acids. These hinder the formation of common secondary structural motifs of proteins usually causing folding of the molecules. There are also no disulfide bridges. As a result, it has relatively little tertiary (three-dimensional) structure. It is relatively hydrophobic, making it poorly soluble in water. It is found in milk as a suspension of particles, called casein micelles.

Whey proteins are a mixture of proteins isolated from whey, and they consist of α-lactalbumin, β-lactoglobulin, serum albumin, and immunoglobulins. The protein in cow's milk is approximately 20% whey and 80% casein. The protein in human milk is approximately 60% whey and 40% casein. There is evidence that whey protein is more bioavailable than casein or soy protein.

Mucins are a family of high molecular weight, heavily glycosylated proteins (glycoconjugates) produced by epithelial tissues in most animals. Mucins' key characteristic is their ability to form gels; therefore they are a key component in most gel-like secretions, serving functions from lubrication to cell signaling to forming chemical barriers. They often take an inhibitory role.

As mentioned, the major categories of proteins vary between mammalian species. Cow's milk is casein predominant, while human milk is whey predominant [13]. Because of this discrepancy, as with several other human milk components, the industry has attempted to match the whey to casein ratio in formulas to represent more closely that found in human milk. Despite this, the types of caseins still differ, and the dynamic fluxes of different compounds found in human milk as well as their interactions have been a major challenge for industry [13].

Casein and most whey proteins in breast milk are utilized for growth. Their concentrations change over the course of lactation: whey protein concentrations during the first 2 weeks of lactation are high, and caseins are low. This results in a whey:casein ratio as high as 80:20. The ratio drops to 65:35 by week 2 and stays constant at about 60:40 thereafter [14]. Some of the formulas used a couple of decades ago contained high casein concentrations. Reports of lactobezoar formation in infants receiving high casein formulas were reported [15,16]. These are hard milk and mucous collections usually found in the stomach that cause intestinal obstruction.

There are also apparent sex differences in protein requirements. Protein gains for boys differ at certain lactational stages from girls over the first 2 years of age [17].

Protein transitions over time

After the first several days post birth during which the first milk, colostrum, is produced, human breast milk composition becomes relatively stable between 2 and 12 weeks of lactation. As the mammary gland matures from a transitional to a mature stage there is less protein variance [18]. The protein concentration over the entire lactation period in human milk is typically in the range of 10–20 g/L [19], starting high in colostrum and then steadily declining until it reaches a plateau and then becomes designated as mature milk (Table 2.1).

Daily protein gain is highest in the very young infant, a time during which relative growth is more rapid. This quickly decreases during later infancy. During the first months, protein gain is three times higher than

Table 2.1 Provides a summary of the major protein concentrations [20].

Protein name	Total	Colostrum protein	Early milk protein	Transitional milk protein	Mature milk protein
Total protein	203–1752	360–1690	606–1675	203–1752	362–1632
Total caseins	19–591	42–507	103–355	87–591	19–743
Ratio whey/casein		90:10	78:22	72:28	60:40

Note: All concentrations are in mg/100 mL.

Source: Adapted from Zhu J, Dingess KA. The functional power of the human milk proteome. Nutrients 2019;11(8): 1834. doi:10.3390/nu11081834

between 12 and 24 months. Thus protein concentration in the mother's milk is accordingly high during the first few weeks of lactation. It continues to diminish throughout the first year, but is much slower than observed in the first weeks [21].

Protein digestion, absorption, and assimilation

Proteins are polymer chains made of amino acids linked together by peptide bonds. Digestion starts in the stomach with the acid/protease hydrolysis, continues in the upper small intestine with pancreatic enzyme activities, through which proteins are broken down to simple amino acids or small peptides, which are subsequently transported and/or absorbed, further processed by the epithelial cell, then assimilated into the bloodstream.

Dynamic changes in protein composition

Despite the reduction in protein concentrations in human milk over time, the nutritional value of protein in breast milk, as measured by the ratio of essential amino acids to total amino acids, is consistent. These changes correlate well with the evolving needs of the growing infant [22].

Low-birth-weight infants have a higher daily protein gain per unit of body weight, thus low-birth-weight infants have higher protein requirements than term infants [23]. Breast milk of mothers who deliver preterm has higher protein and amino acid concentrations than breast milk of mothers who deliver at term [22]. In very preterm infants, feeding breast milk without supplements does not meet protein requirements. Supplements which are on the market are based on protein fractions of cow's milk or human donor milk. Supplementation of breast milk or the use of preterm formulas improves growth rates of low-birth-weight infants. This is supported by multiple studies that show improved weight and height gain as well as helping in the prevention of metabolic bone disease [24].

Lipids in human milk

Lipids are a major source of energy provided by human milk to the infant [25,26]. They are comprised in human milk largely in the form of triglycerides. Some of the lipids are polyunsaturated fatty acids (PUFAs) with 18 carbon atoms on the lipid chains that are attached to the glycerol molecule. Of these, linoleic and linolenic acids are considered as essential lipids, which cannot be synthesized in the body. They also contain lipid-soluble vitamins. In addition to being an important energy source, lipids also exert several biological effects on gastrointestinal function, membrane composition and function, infant growth, neurodevelopment, and immune function [27].

Human milk lipids

Human milk lipids supply a major portion of the total energy intake in young infants (mean = 44% of the energy supply) [1]. The average intake of human milk lipids in fully breastfed infants amounts to 21–42 g/day between birth and 6 months [1]. Among the macronutrients in milk, fat shows the most variable concentration. Milk fat content tends to increase with a longer duration of breastfeeding. Milk fat also varies during different parts of the day [28]. Milk fat concentration increases with maternal fat deposition during pregnancy. It also increases when there is an increasing time interval to the preceding milk expression from the same breast [29].

Milk fat increases during each breastfeeding meal, with markedly higher milk fat contents in hindmilk (at the end of feeding) than in foremilk (at the beginning of the feed [30,31]). This may provide a benefit in which infants will initially get milk rich in the essential water-soluble substrates. Those who are hungrier and drink more milk obtain more fat and energy to satisfy their caloric needs.

Lipid nomenclature

To have a better understanding of lipids in human milk and their physiology, a brief review of the nomenclature is in order. Most dietary lipids are in the form of triglycerides, sometimes also called triacylglycerols. They are composed of a glycerol backbone and fatty acids are attached to the three carbon atoms of this backbone. The length of the fatty acids determines whether they are short, middle, or long chains. The number of double bonds between the carbon atoms determines the degree of unsaturation. The essential lipids, linoleic and linolenic, are metabolic precursors to the long-chain fatty acids.

Milk fat composition

Milk fat is an emulsion of fat globules in an aqueous liquid [27]. These globules vary widely in size and are formed by the alveolar cells. Fig. 2.1 below shows a milk fat globule as it is secreted by the mammary epithelial cell.

In the core of the globule, triacylglycerols are the main constituent. The triglyceride-rich core is derived from the endoplasmic reticulum of the maternal epithelial cell. It consists of a monolayer primarily of

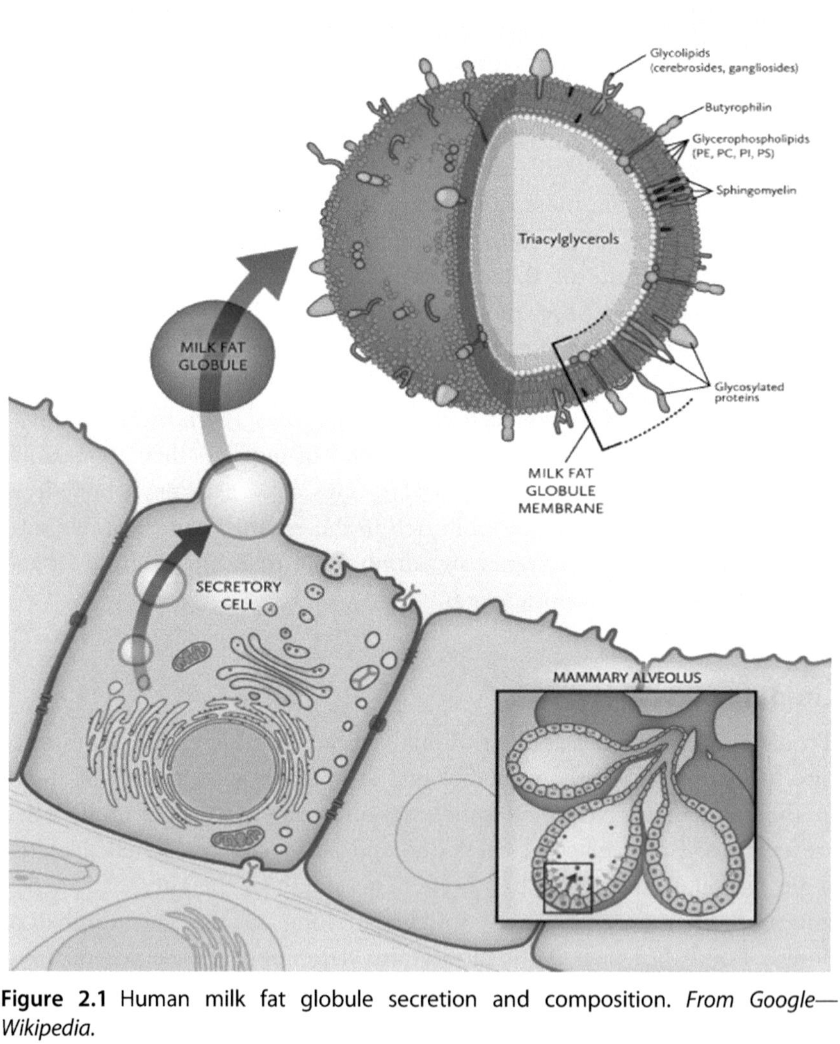

Figure 2.1 Human milk fat globule secretion and composition. *From Google—Wikipedia.*

phosphatidylethanolamine, phosphatidylserine, phosphatidylinositol, and cholesterol.

From several studies, it has been concluded that the milk fat globule structure may have important biological roles for development of the nervous and immune systems. These observations lead to the conclusion that the milk fat globule membrane plays important biological roles for the development of nervous and immune functions.

Cholesterol

Milk fat globule lipids also provide considerable amounts of free and esterified cholesterol. Human milk contains considerably more cholesterol than formula: (cholesterol content of 90–150 mg/L in human milk in contrast to typically only 0–4 mg/L in infant formula). Although this seems paradoxical when considering the role of cholesterol in atherogenesis, cholesterol is a critical building block for cell membranes. It is incorporated in high quantities into myelin during the period of rapid brain growth. Cholesterol is a critical substrate for the synthesis of bile acids, lipoproteins, vitamin D, hormones, and oxysterols that modulate cholesterol, lipid, and glucose homeostasis [32,33]. The provision of preformed cholesterol also appears to serve a feedback mechanism: there is about a threefold lower endogenous cholesterol synthesis rate in breastfed than formula-fed infants [34].

Fatty acids provided with milk lipids

Triacylglycerols contribute some 98%–99% of human milk fat. The properties of milk triglycerides are highly influenced by their fatty acid composition. About 70% of human milk palmitic acid is esterified in the middle position (sn-2 position) of triacylglycerols. This is thought to improve absorption. During digestion, fatty acids are liberated as nonesterified fatty acids by pancreatic lipases. These nonesterified fatty acids are quite well absorbed. However, liberated long-chain saturated fatty acids, such as palmitic acid, are poorly water soluble and poorly absorbed. These bind to calcium and form calcium soaps that are excreted with stools. This results in lower fat and calcium absorption. However, if palmitic acid is esterified in the sn-2 position, which it is for the most part in human milk lipids, pancreatic breakdown of fats is to palmitoyl-monoglycerol which is water soluble, well absorbed, and reduces fat and calcium malabsorption [35].

Long-chain polyunsaturated fatty acids

The provision of arachidonic acid and docosahexaenoic acid (DHA) has received considerable attention. This is based on the numerous biological effects of these essential fatty acids in early life. These effects appear to be mediated by these longer-chain fatty acids rather than the precursor essential fatty acids.

The human milk contents of the monounsaturated fatty acid oleic acid (nonessential) and of the essential PUFA linoleic acid and α-linolenic acid depend on the maternal dietary intake of these fatty acids. The DHA intake of lactating women determines the DHA content in their breast milk [36].

Providing DHA enhances the early development of visual acuity. The European Food Safety Authority concluded that a cause-and-effect relationship has been established between the intake of infant and follow-on formula supplemented with DHA to visual function at 12 months in formula-fed infants born at term from birth up to 12 months. Similar findings are seen in breastfed infants after weaning up to 12 months. However, there still is remaining controversy pertaining to the effects of the supply of preformed LC-PUFA on neurodevelopment of healthy term infants [37].

In infants born preterm, LC-PUFAs are likely to play a very critical role. During pregnancy, the fetus derives considerable lipid including preformed LC PUFA from the mother via the placenta (Fig. 2.2, left). This is suddenly interrupted if the infant is born preterm and received an intravenous lipid infusion that contains sparse LC PUFA and is not taking full enteral feedings (Fig. 2.2, left) [38]. Provision of mother's milk to these infants may at least partially ameliorate this deficiency because of the high concentrations of these fatty acids. However, this is dependent on the maternal diet. When high in certain types of fish, these fatty acids will be seen in higher concentrations in the milk. They may be responsible for certain beneficial effects for the infants [39].

Micronutrients

Micronutrients (vitamins and minerals) play a critical role in sustaining the metabolic demands and contribute directly to the maturation of the central nervous system. Processes that relate to generation of energy require glucose, iron, copper, zinc, and selenium. Neural cell differentiation is supported by iodine and zinc. Myelination is affected by copper, iodine, vitamin B12, choline, and cholesterol. The structure and

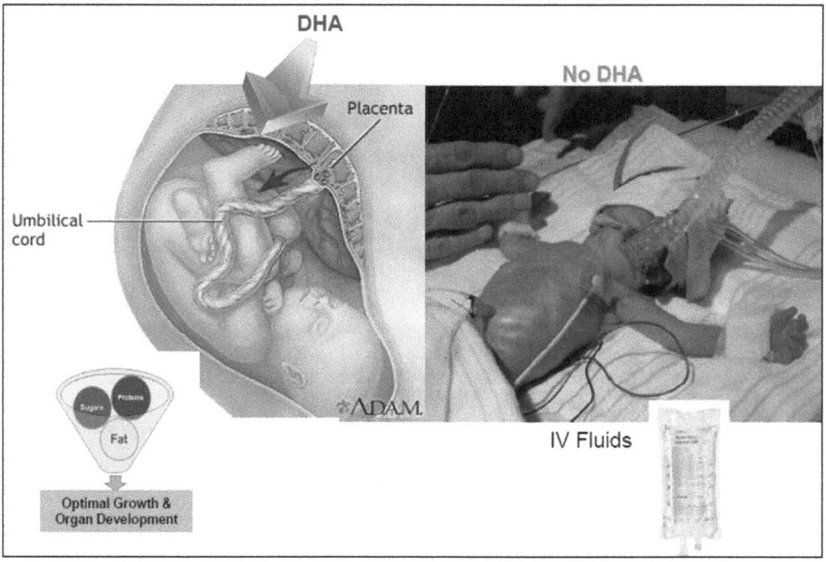

Figure 2.2 Long-chain polyunsaturated fatty acids in mother's milk.

composition of the brain is affected by folate. There is a wealth of literature and experience with deficiencies of these micronutrients [40]. In fact, approximately two billion people a year suffer from micronutrient deficiencies contributing to worldwide morbidity and mortality.

These micronutrients are thought to be present in sufficient quantities in human milk for most infants of healthy, well-nourished mothers. However, the micronutrient composition of human milk varies across populations is this thought to be sensitive to the dietary intake of mothers. The milk concentration of many of these is closely associated with maternal levels. For example, if a mother is deficient in iodine, B1, B2, B6, B12, vitamin A, and selenium, this can be reflected in her milk and the exclusively breastfed infant is also at high risk of becoming deficient.

Unfortunately, the number of studies that relate breast milk micronutrient levels to health outcomes is few, and more research in this area is needed.

Summary

The macronutrient composition of human milk is especially suited for the human infant, and in turn infants are provided with major energy and protein synthetic needs as well as essential fatty acids. The composition of these macronutrients is dynamic, with the largest changes occurring early during the lactation period. Less is known about micronutrients, but their concentrations in breast milk reflect the mother's nutritional status, and this likely is also reflected in the infant's micronutrient nutritional status.

References

[1] Grote V, Verduci E, Scaglioni S, et al. Breast milk composition and infant nutrient intakes during the first 12 months of life. Eur J Clin Nutr 2016;70(2):250–6. Available from: https://doi.org/10.1038/ejcn.2015.162.

[2] Indrio F, Neu J, Pettoello-Mantovani M, et al. Development of the gastrointestinal tract in newborns as a challenge for an appropriate nutrition: a narrative review. Nutrients 2022;14(7):1405. Available from: https://doi.org/10.3390/nu14071405.

[3] Lenfestey MW, Neu J. Gastrointestinal development: implications for management of preterm and term infants. Gastroenterol Clin North Am 2018;47(4):773–91. Available from: https://doi.org/10.1016/j.gtc.2018.07.005.

[4] Neu J, Koldovsky O. Nutrient absorption in the preterm neonate. Clin Perinatol 1996;23(2):229–43.

[5] Bode L. 1.4.6 Oligosaccharides in human milk. World Rev Nutr Diet 2022;124:115–21. Available from: https://doi.org/10.1159/000518201.

[6] Berger PK, Fields DA, Demerath EW, Fujiwara H, Goran MI. High-fructose corn-syrup-sweetened beverage intake increases 5-hour breast milk fructose concentrations in lactating women. Nutrients 2018;10(6):669. Available from: https://doi.org/10.3390/nu10060669.

[7] Goran MI, Martin AA, Alderete TL, Fujiwara H, Fields DA. Fructose in breast milk is positively associated with infant body composition at 6 months of age. Nutrients 2017;9(2):146. Available from: https://doi.org/10.3390/nu9020146.

[8] Berger PK, Plows JF, Jones RB, et al. Associations of maternal fructose and sugar-sweetened beverage and juice intake during lactation with infant neurodevelopmental outcomes at 24 months. Am J Clin Nutr 2020;112(6):1516–22. Available from: https://doi.org/10.1093/ajcn/nqaa255.

[9] Fields DA, George B, Williams M, et al. Associations between human breast milk hormones and adipocytokines and infant growth and body composition in the first 6 months of life. Pediatr Obes 2017;12(Suppl 1):78–85. Available from: https://doi.org/10.1111/ijpo.12182.

[10] Gridneva Z, Rea A, Tie WJ, et al. Carbohydrates in human milk and body composition of term infants during the first 12 months of lactation. Nutrients 2019;11(7):1472. Available from: https://doi.org/10.3390/nu11071472.

[11] Larsson MW, Lind MV, Laursen RP, et al. Human milk oligosaccharide composition is associated with excessive weight gain during exclusive breastfeeding – an explorative study. Front Pediatr 2019;7:297. Available from: https://doi.org/10.3389/fped.2019.00297.

[12] Berger PK, Plows JF, Jones RB, et al. Human milk oligosaccharide 2'-fucosyllactose links feedings at 1 month to cognitive development at 24 months in infants of normal and overweight mothers. PLoS One 2020;15(2):e0228323. Available from: https://doi.org/10.1371/journal.pone.0228323.

[13] Haschke F, Haiden N, Thakkar SK. Nutritive and bioactive proteins in breastmilk. Ann Nutr Metab 2016;69(Suppl 2):17–26. Available from: https://doi.org/10.1159/000452820.
[14] Lönnerdal B, Erdmann P, Thakkar SK, Sauser J, Destaillats F. Longitudinal evolution of true protein, amino acids and bioactive proteins in breast milk: a developmental perspective. J Nutr Biochem 2017;41:1–11. Available from: https://doi.org/10.1016/j.jnutbio.2016.06.001.
[15] Heinz-Erian P, Gassner I, Klein-Franke A, et al. Gastric lactobezoar - a rare disorder? Orphanet J Rare Dis 2012;7:3. Available from: https://doi.org/10.1186/1750-1172-7-3.
[16] Castro L, Berenguer A, Pilar C, Gonçalves R, Nunes JL. Recurrent gastric lactobezoar in an infant. Oxf Med Case Rep 2014;2014(4):80–2. Available from: https://doi.org/10.1093/omcr/omu031.
[17] Fomon SJ, Haschke F, Ziegler EE, Nelson SE. Body composition of reference children from birth to age 10 years. Am J Clin Nutr 1982;35(5 Suppl):1169–75. Available from: https://doi.org/10.1093/ajcn/35.5.1169.
[18] Gidrewicz DA, Fenton TR. A systematic review and meta-analysis of the nutrient content of preterm and term breast milk. BMC Pediatr 2014;14:216. Available from: https://doi.org/10.1186/1471-2431-14-216.
[19] Donovan S.M. Human milk proteins: composition and physiological significance. In: Human milk: composition, clinical benefits and future opportunities. Karger Publishers; 2019. p. 93–101.
[20] Zhu J, Dingess KA. The functional power of the human milk proteome. Nutrients 2019;11(8):1834. Available from: https://doi.org/10.3390/nu11081834.
[21] Haschke F, Haiden N, Thakkar SK. Nutritive and bioactive proteins in breastmilk. Ann Nutr Metab 2016;69(Suppl. 2):16–26. Available from: https://doi.org/10.1159/000452820.
[22] Zhang Z, Adelman AS, Rai D, Boettcher J, Lőnnerdal B. Amino acid profiles in term and preterm human milk through lactation: a systematic review. Nutrients 2013;5(12):4800–21. Available from: https://doi.org/10.3390/nu5124800.
[23] Ziegler EE, O'Donnell AM, Nelson SE, Fomon SJ. Body composition of the reference fetus. Growth 1976;40(4):329–41.
[24] O'Connor DL, Jacobs J, Hall R, et al. Growth and development of premature infants fed predominantly human milk, predominantly premature infant formula, or a combination of human milk and premature formula. J Pediatr Gastroenterol Nutr 2003;37(4):437–46. Available from: https://doi.org/10.1097/00005176-200310000-00008.
[25] Koletzko B, Agostoni C, Bergmann R, Ritzenthaler K, Shamir R. Physiological aspects of human milk lipids and implications for infant feeding: a workshop report. Acta Paediatr (Oslo, Norway: 1992) 2011;100(11):1405–15. Available from: https://doi.org/10.1111/j.1651-2227.2011.02343.x.
[26] Delplanque B, Gibson R, Koletzko B, Lapillonne A, Strandvik B. Lipid quality in infant nutrition: current knowledge and future opportunities. J Pediatr Gastroenterol Nutr 2015;61(1):8–17. Available from: https://doi.org/10.1097/mpg.0000000000000818.
[27] Koletzko B. Human milk lipids. Ann Nutr Metab 2016;69(Suppl 2):28–40. Available from: https://doi.org/10.1159/000452819.
[28] Andreas NJ, Kampmann B, Mehring Le-Doare K. Human breast milk: a review on its composition and bioactivity. Early Hum Dev 2015;91(11):629–35. Available from: https://doi.org/10.1016/j.earlhumdev.2015.08.013.
[29] Michaelsen KF, Larsen PS, Thomsen BL, Samuelson G. The Copenhagen cohort study on infant nutrition and growth: breast-milk intake, human milk macronutrient

content, and influencing factors. Am J Clin Nutr 1994;59(3):600–11. Available from: https://doi.org/10.1093/ajcn/59.3.600.
[30] Takumi H, Kato K, Nakanishi H, et al. Comprehensive analysis of lipid composition in human foremilk and hindmilk. J Oleo Sci 2022;71(7):947–57. Available from: https://doi.org/10.5650/jos.ess21449.
[31] Khan S, Hepworth AR, Prime DK, Lai CT, Trengove NJ, Hartmann PE. Variation in fat, lactose, and protein composition in breast milk over 24 hours: associations with infant feeding patterns. J Hum Lactation 2013;29(1):81–9. Available from: https://doi.org/10.1177/0890334412448841.
[32] Cartocci V, Servadio M, Trezza V, Pallottini V. Can cholesterol metabolism modulation affect brain function and behavior? J Cell Physiol 2017;232(2):281–6. Available from: https://doi.org/10.1002/jcp.25488.
[33] Mutemberezi V, Guillemot-Legris O, Muccioli GG. Oxysterols: from cholesterol metabolites to key mediators. Prog Lipid Res 2016;64:152–69. Available from: https://doi.org/10.1016/j.plipres.2016.09.002.
[34] Wong WW, Hachey DL, Insull W, Opekun AR, Klein PD. Effect of dietary cholesterol on cholesterol synthesis in breast-fed and formula-fed infants. J Lipid Res 1993;34(8):1403–11.
[35] Carnielli VP, Luijendijk IH, Van Goudoever JB, et al. Structural position and amount of palmitic acid in infant formulas: effects on fat, fatty acid, and mineral balance. J Pediatr Gastroenterol Nutr 1996;23(5):553–60. Available from: https://doi.org/10.1097/00005176-199612000-00007.
[36] Fidler N, Sauerwald T, Pohl A, Demmelmair H, Koletzko B. Docosahexaenoic acid transfer into human milk after dietary supplementation: a randomized clinical trial. J Lipid Res 2000;41(9):1376–83.
[37] Qawasmi A, Landeros-Weisenberger A, Bloch MH. Meta-analysis of LCPUFA supplementation of infant formula and visual acuity. Pediatrics 2013;131(1):e262–72. Available from: https://doi.org/10.1542/peds.2012-0517.
[38] Lapillonne A, Moltu SJ. Long-chain polyunsaturated fatty acids and clinical outcomes of preterm infants. Ann Nutr Metab 2016;69(Suppl 1):35–44. Available from: https://doi.org/10.1159/000448265.
[39] Lauritzen L, Carlson SE. Maternal fatty acid status during pregnancy and lactation and relation to newborn and infant status. Matern Child Nutr 2011;7(Suppl 2):41–58. Available from: https://doi.org/10.1111/j.1740-8709.2011.00303.x.
[40] Domellöf M. Micronutrient intakes and health outcomes in preterm infants. Nestle Nutr Inst Workshop Ser 2021;96:130–7. Available from: https://doi.org/10.1159/000519393.

CHAPTER THREE

Bioactive components of human milk

"The doctor of the future will no longer treat the human frame with drugs, but rather will cure and prevent disease with nutrition."

Thomas Edison

Key Points

1. Bioactive components are not the basic macro or micronutrients required for metabolic needs but include activities such as immunoprotection, homeostasis, and digestion.
2. To provide bioactive components in commercial formulas, they need to be added. This poses numerous challenges for the industry trying to mimic these compositions.

Introduction

Overview

Numerous components of human milk are considered "bioactive." These are distinguished from the macro and micronutrients in which they are not utilized directly for nutritional purposes but rather are adapted for specialized needs of the infant such as immunoprotection and immunomodulation, metabolic homeostasis, digestion, and regulatory activities. These factors are usually not found in commercial formulas unless specifically added. This chapter summarizes a few of the known bioactive components found in human milk. Others such as oligosaccharides, immunoglobulins, and microbes will be covered in detail in subsequent chapters.

The accompanying Fig. 3.1 illustrates some of the bioactive components of human milk that are involved in health outcomes and their

The Science of Human Milk.
DOI: https://doi.org/10.1016/B978-0-323-99913-7.00003-4

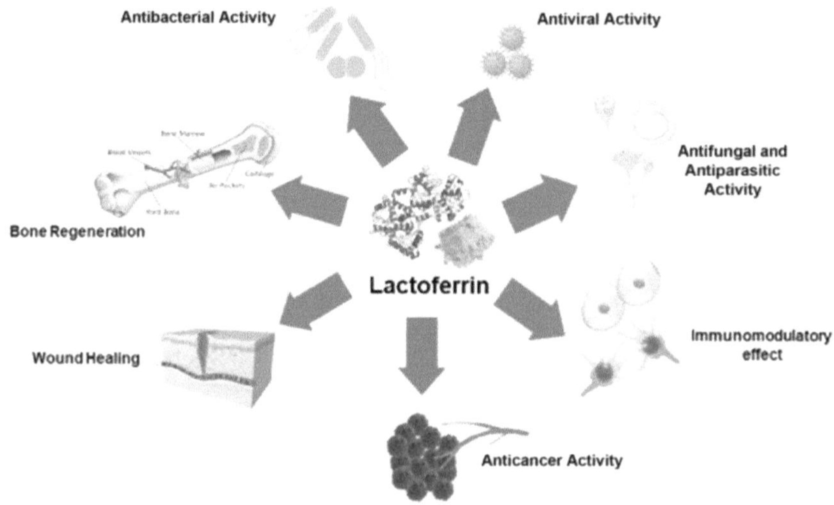

Figure 3.1 Multifaceted role of lactoferrin from human milk. *From https://www.sciencedirect.com/science/article/pii/S0142961220306013.*

functions [1,2]. It should be noted that many of these compounds are targets of active research for addition to commercial formulas because of their potentially beneficial activities.

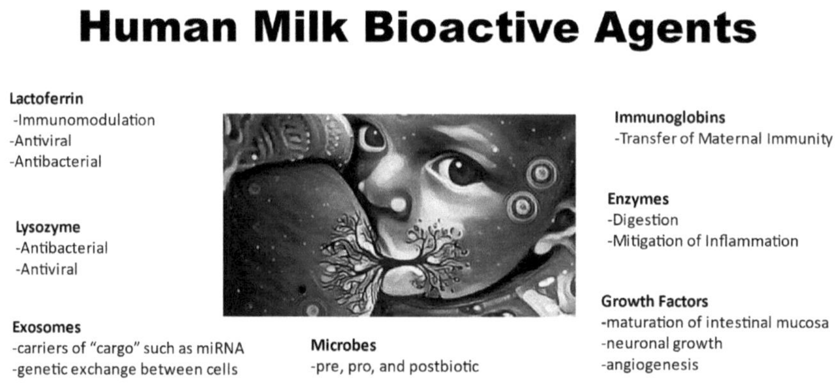

Here we will describe a few of these.

Lactoferrin

In the early 1970s, breast milk was found to have bacteriostatic activity against *Escherichia coli* [3]. The active agent was found to be an iron-binding protein with a high affinity for iron which enables its ability to withhold iron from iron-requiring pathogens. This agent, lactoferrin, has been found to play many additional roles.

Lactoferrin is the second most abundant protein in breast milk. Its role is multifaceted (Fig. 3.1), playing a major role in immunity of the infant, preventing infection, playing a role in iron metabolism, having antiinflammatory properties, and being an antioxidant.

One of the main roles of lactoferrin is to bind with and transport iron in the body. It binds iron and is transferred via receptors into and between cells, serum, bile, and cerebrospinal fluid. Human milk lactoferrin levels change as milk matures with colostrum having higher concentrations in both term and preterm milk [4,5]; however, preterm milk tends to maintain higher levels of lactoferrin over time [5]. A recent study of Chinese women reported that lactoferrin concentration was 3.16 and 1.73 g/L in colostrum and milk, respectively [6]. LF binds to free iron, which is an essential nutrient for bacterial growth. This confers a bacteriostatic effect, which inhibits the growth or reproduction of bacteria. Furthermore, the growth of those bacteria that do not require much iron is therefore enhanced. Many of these are beneficial to humans and include *Lactobacillus* and *Bifidobacterium*. Lactoferrin is present in cow's milk, but studies in vitro have shown that human milk was more effective in preventing bacterial growth than cow's milk.

Lactoferrin (LF) has important immunological properties and is both antibacterial and antiviral. Highly relevant to the recent pandemic is evidence that LF can bind to at least some of the receptors used by coronaviruses and thereby block their entry. Lactoferrin may eventually be of preventive and therapeutic value during the present COVID-19 pandemic.

Lactoferrin is also able to directly kill bacteria, that is, it has a bactericidal function. With its high positive ionic charge, lactoferrin can form a complex with lipopolysaccharide (LPS), a bacterial cell wall component, which is highly negatively charged and create holes in the outer membrane of Gram-negative bacteria that contain high quantities of the LPS. These holes are penetrated by another protein, lysozyme, which degrades the proteoglycan matrix and kills the bacterium [7].

Several studies have investigated the addition of cow's milk lactoferrin to breast milk, donor human milk, and formulas. No differences were seen in the late outcomes of these preterm infants between treated and control infants [8–10]. Another study showed that bovine lactoferrin added to infants given oral supplements of lactoferrin showed a significant reduction in sepsis in premature infants [11].

Lysozyme

Lysozyme was first described by Alexander Fleming in 1922 [12]. It is an enzyme that catalyzes the hydrolysis of 1,4 beta linkages between sugar molecules found in peptidoglycan (Fig. 3.2), the major component of the Gram-positive bacterial cell wall.

Lysozyme is widely distributed in tears, saliva, milk [13,14] and plays important roles as a nonspecific immune factor and antiinflammatory factor [15]. It increases the levels of beneficial intestinal microbes, thus improving disease resistance [16,17]. The content of lysozyme in human milk ranges from 3 to 3000 μg/mL, and the typical concentration is about 200–400 μg/mL [18,19]. Cow's milk typically contains only 0.05–0.22 μg/mL of lysozyme [20]. In addition, its activity is 1/10 of lysozyme from human breast milk [21]. It is important to note that despite the presence of some of these bioactive components in cow's and nonhuman species milk, these are usually treated in a manner (pasteurization, defatting, etc.) where these bioactive components are lost.

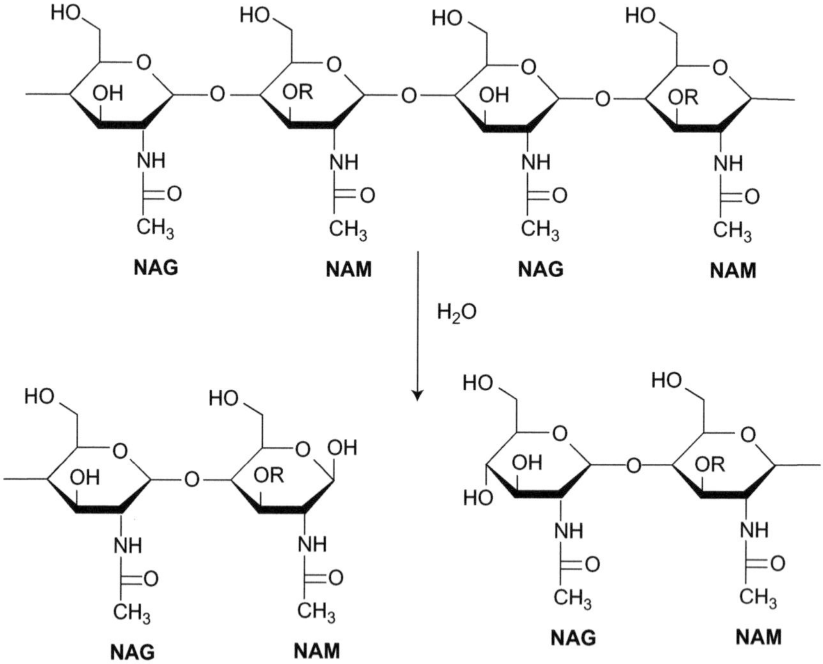

Figure 3.2 Breakdown of peptidoglycan by lysozyme.

Lysozyme hydrolyzes a major bacterial cell wall protein-carbohydrate component called peptidoglycan, which is the major component of Gram-positive bacterial cell wall. This consequently causes hydrolysis, in turn, compromises the integrity of bacterial cell walls causing lysis (or disintegration) of the bacteria, hence its name. Lysozyme is also abundant in other secretions including tears, saliva, and mucus.

Lysozyme role in disease and therapy

Lysozyme is part of the innate immune system. Reduced lysozyme levels have been associated with bronchopulmonary dysplasia (BPD) in newborns [22]. In this study, tracheal aspirate samples were collected from 36 newborn infants with respiratory distress syndrome who underwent intubation during days 1−28 of life and analyzed for the mucosal antimicrobial lysozyme. For babies in whom BPD developed, lysozyme levels in the secretions of babies were also lower on day 3 (31 ± 5 micrograms/mL) than in those for babies who recovered (54 ± 7.5 micrograms/mL). For babies with BPD whose endotracheal tube remained in place beyond day 4, lysozyme levels on days 5−12 were significantly lower for those classified as having severe BPD than for those with mild to moderate BPD. The relevance of this finding to lysozyme in human milk is not yet clear but shows an association between a lysozyme, which is found in high concentrations in human milk and a lower incidence of a major problem seen in preterm infants.

Bile salt-stimulated lipase

Lipid digestion is critical for infant development. Milk fat is partially hydrolyzed in the stomach and further hydrolyzed in the upper small intestine. The major lipases aiding fat digestion in the newborn small intestine are therefore pancreatic lipase and bile salt-stimulated lipase (BSSL) derived from the exocrine pancreas and milk, respectively [23]. Fig. 3.3 summarizes some of these activities.

BSSL catalyzes the digestion of triglycerides in human milk fat and uniquely breaks the bond of palmitic acid and glycerol in the sn-2 position, a common property of human milk fat. Many other lipases cannot break this bond, which seems to explain the high efficiency of lipid digestion in breast-fed infants [24]. There is no BSSL in cow's milk and there is considerable interest in using human recombinant forms of this enzyme to aid in lipid digestion in infants not fed their mother's milk. The activity

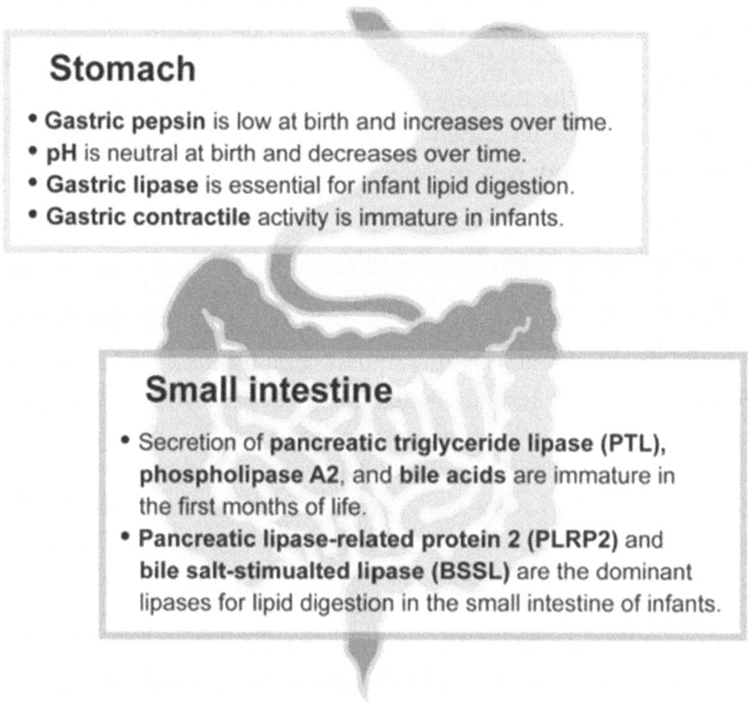

Figure 3.3 Intestinal lipid digestive activities.

of this enzyme is largely lost with usual pasteurization processes in the processing of donor human milk [25].

Alkaline phosphatase in human milk

Alkaline phosphatase (ALP) is an enzyme found in high concentrations in human milk that is almost completely degraded by the pasteurization process. Breast-fed infants demonstrate higher fecal ALP activity, which is known to dephosphorylate and detoxify the endotoxin component of LPS, a known pathogenic component of numerous bacteria that is capable of initiating inflammatory responses. It has been proposed that the effect of human milk on the neonatal activity of ALP is not only due to an intake of exogenous ALP but also to an induction of the enzyme by some agent in human milk, and the agent is absent in neonatal formula.

It is not known whether this bioactive component of human milk plays a role in protection of human infants against disease and requires additional study.

Human milk extracellular vesicles and exosomes

These will be briefly covered here but described in greater detail in Chapter 8, Human Milk Exosomes/Extracellular Vesicles. Extracellular vesicles describe different subsets which include exosomes and microvesicles [26]. These are differentiated from one another using methods such as ultracentrifugation. These act as carriers containing various molecules such as proteins, microRNAs (miRNAs), and metabolites.

miRNAs are small noncoding RNA (~22 nucleotides) that regulate the posttranscriptional expression of genes, which in turn exhibit biological activities in humans [27,28]. Human milk contains several miRNAs [29], which because of their packaging in exosomes have the ability to survive in the acidic environment in the intestine and can also be absorbed by the intestinal epithelium [30]. Infant formulas have lower concentrations of miRNAs compared with human milk [31]. It is thought that these miRNAs are likely derived from immune-related and mammary gland cells [32]. The studies suggest that milk miRNAs are taken up by intestinal, immune, and cancer cell lines [33].

Several studies suggest that these miRNAs are likely impacting the immune system. Some of these are thought to regulate the innate immune response [34], TLR4 signaling, and macrophage activation [35]. We still do not understand the specific roles of these milk miRNAs and additional studies are needed to better define these roles. In addition, whether miRNAs have direct or indirect effects via microbiota on the infant gut and the immune system is not fully understood. Some evidence suggests that miRNA could change microbiota composition.

Another role for exosomes-mediated miRNA is a novel mechanism of genetic exchange (cargo delivery) between maternal mammary epithelium and infant cells. It is therefore possible that exosomes in milk may survive digestion and deliver miRNAs to intestinal cells, or, if transferred into the bloodstream, to cells in other tissues. The transfer of this genetic material to the infant can affect gene transcription and regulation of cellular events in several tissues [36].

Growth factors and hormones in human milk

Several hormones and growth factors are found in human milk that serve as bioactive agents [37]. Examples of these are seen in the table below. Many of their effects as they relate to breastfeeding and health outcomes in infants are not yet well known. Others have been more extensively studied (Table 3, below—adapted from Reference [37]) [38,39].

Epidermal growth factors' levels are 2000 times higher in colostrum than in mature milk and decrease with lactation. They appear to play a critical role in intestinal maturation and repair [38,40]. Insulin-like growth factor (IGF)-1 and IGF-2 are also abundant in colostrum and decrease with lactation. IGF is taken up in its bioactive form by the intestine and transported to the blood [41]. IGF-1 protects the intestinal epithelium against intestinal damage caused by oxidative stress; furthermore, it stimulates erythropoiesis and helps increase hematocrit [42].

Brain-derived neurotrophic factor acts on the enteric nervous system and is necessary for the development of immature intestine in infants [43]. These neuronal growth factors are found in human breast milk for up to 90 days after birth [44].

Vascular endothelial growth factor and its antagonists are thought to help regulate angiogenesis [45]. Whether these factors in breast milk play a role in the prevention in retinopathy of prematurity is unknown.

Erythropoietin stimulates red blood cell production and is thought to help prevent anemia of prematurity [46] (Table 3.1).

Table 3.1 Growth factors in human breast milk and their functions.

Growth factors	Functions
Epidermal growth factor	Maturation and healing of the intestinal mucosa
Neuronal growth factors	Development of the enteric nervous system in newborns
Brain-derived neurotrophic	Alleviation of impaired peristalsis
Cell-line-derived neurotrophic	Neuron survival and outgrowth
Insulin-like growth factor	Stimulation of erythropoiesis
Vascular endothelial growth	Regulation of angiogenesis
Erythropoietin	Responsible for increasing red blood cells
Adiponectin	Regulation of metabolism and suppression of inflammation

Summary

This chapter has briefly described some of the bioactive components found in human breast milk that play roles well beyond those provided by the macronutrients. At this juncture, many of these are under active investigation as additives for commercial formulas. Despite this, it will be very difficult to mimic their actions as found in human breast milk, largely because of the myriad of bioactive factors present, their interactions, the unique matrix of human milk, and its dynamic nature. All of these will be discussed further in subsequent chapters.

References

[1] Lönnerdal B. Bioactive proteins in human milk: health, nutrition, and implications for infant formulas. J Pediatr 2016;173:S4–9. Available from: https://doi.org/10.1016/j.jpeds.2016.02.070.

[2] Lönnerdal B. Bioactive proteins in human milk: mechanisms of action. J Pediatr 2010;156(2 Suppl):S26–30. Available from: https://doi.org/10.1016/j.jpeds.2009.11.017.

[3] Bullen JJ. Iron-binding proteins and other factors in milk responsible for resistance to Escherichia coli. Ciba Found Symp 1976;42:149–69. Available from: https://doi.org/10.1002/9780470720240.ch9.

[4] Hirai Y, Kawakata N, Satoh K, et al. Concentrations of lactoferrin and iron in human milk at different stages of lactation. J Nutr Sci Vitaminol (Tokyo) 1990;36(6):531–44. Available from: https://doi.org/10.3177/jnsv.36.531.

[5] Albenzio M, Santillo A, Stolfi I, et al. Lactoferrin levels in human milk after preterm and term delivery. Am J Perinatol 2016;33(11):1085–9. Available from: https://doi.org/10.1055/s-0036-1586105.

[6] Yang Z, Jiang R, Chen Q, et al. Concentration of lactoferrin in human milk and its variation during lactation in different Chinese populations. Nutrients 2018;10(9):1235. Available from: https://doi.org/10.3390/nu10091235.

[7] Ellison 3rd RT, Giehl TJ. Killing of Gram-negative bacteria by lactoferrin and lysozyme. J Clin Invest 1991;88(4):1080–91. Available from: https://doi.org/10.1172/jci115407.

[8] Griffiths J, Jenkins P, Vargova M, et al. Enteral lactoferrin to prevent infection for very preterm infants: the ELFIN RCT. Health Technol Assess 2018;22(74):1–60. Available from: https://doi.org/10.3310/hta22740.

[9] Griffiths J, Jenkins P, Vargova M, et al. Enteral lactoferrin supplementation for very preterm infants: a randomised placebo-controlled trial. Lancet 2019;393(10170):423–33.

[10] Asztalos EV, Barrington K, Lodha A, Tarnow-Mordi W, Martin A. Lactoferrin infant feeding trial_Canada (LIFT_Canada): protocol for a randomized trial of adding lactoferrin to feeds of very-low-birth-weight preterm infants. BMC Pediatr 2020;20(1):40. Available from: https://doi.org/10.1186/s12887-020-1938-0.

[11] Manzoni P, Rinaldi M, Cattani S, et al. Bovine lactoferrin supplementation for prevention of late-onset sepsis in very low-birth-weight neonates: a randomized trial. JAMA 2009;302(13):1421–8. Available from: https://doi.org/10.1001/jama.2009.1403.

[12] Fleming A. On a remarkable bacteriolytic element found in tissues and secretions. Proc R Soc Lond Ser B 1922;93(653):306–17. Available from: http://www.jstor.org/stable/80959.
[13] Mason DY, Taylor CR. The distribution of muramidase (lysozyme) in human tissues. J Clin Pathol 1975;28(2):124–32. Available from: https://doi.org/10.1136/jcp.28.2.124.
[14] Hankiewicz J, Swierczek E. Lysozyme in human body fluids. Clin Chim Acta 1974;57(3):205–9. Available from: https://doi.org/10.1016/0009-8981(74)90398-2.
[15] Ogundele MO. A novel anti-inflammatory activity of lysozyme: modulation of serum complement activation. Mediators Inflamm 1998;7(5):363–5. Available from: https://doi.org/10.1080/09629359890893.
[16] Lönnerdal B. Nutritional and physiologic significance of human milk proteins. Am J Clin Nutr 2003;77(6):1537s–43s. Available from: https://doi.org/10.1093/ajcn/77.6.1537S.
[17] Levy J. Immunonutrition: the pediatric experience. Nutrition 1998;14(7–8):641–7. Available from: https://doi.org/10.1016/s0899-9007(98)00007-0.
[18] Chandan RC, Shahani KM, Holly RG. Lysozyme content of human milk. Nature 1964;204:76–7. Available from: https://doi.org/10.1038/204076a0.
[19] Montagne P, Cuillière ML, Molé C, Béné MC, Faure G. Changes in lactoferrin and lysozyme levels in human milk during the first twelve weeks of lactation. Adv Exp Med Biol 2001;501:241–7. Available from: https://doi.org/10.1007/978-1-4615-1371-1_30.
[20] Piccinini R, Binda E, Belotti M, Casirani G, Zecconi A. Comparison of blood and milk non-specific immune parameters in heifers after calving in relation to udder health. Vet Res 2005;36(5–6):747–57. Available from: https://doi.org/10.1051/vetres:2005030.
[21] Ito Y, Yamada H, Nakamura M, Yoshikawa A, Ueda T, Imoto T. The primary structures and properties of non-stomach lysozymes of sheep and cow, and implication for functional divergence of lysozyme. Eur J Biochem 1993;213(2):649–58. Available from: https://doi.org/10.1111/j.1432-1033.1993.tb17805.x.
[22] Revenis ME, Kaliner MA. Lactoferrin and lysozyme deficiency in airway secretions: association with the development of bronchopulmonary dysplasia. J Pediatr 1992;121(2):262–70. Available from: https://doi.org/10.1016/s0022-3476(05)81201-6.
[23] He X, McClorry S, Hernell O, Lönnerdal B, Slupsky CM. Digestion of human milk fat in healthy infants. Nutr Res 2020;83:15–29. Available from: https://doi.org/10.1016/j.nutres.2020.08.002.
[24] Lombardo D. Bile salt-dependent lipase: its pathophysiological implications. Biochim Biophys Acta 2001;1533(1):1–28. Available from: https://doi.org/10.1016/s1388-1981(01)00130-5.
[25] Koh J, Victor AF, Howell ML, et al. Bile salt-stimulated lipase activity in donor breast milk influenced by pasteurization techniques. Front Nutr 2020;7:552362. Available from: https://doi.org/10.3389/fnut.2020.552362.
[26] O'Reilly D, Dorodnykh D, Avdeenko NV, et al. Perspective: the role of human breast-milk extracellular vesicles in child health and disease. Adv Nutr (Bethesda, MD) 2021;12(1):59–70. Available from: https://doi.org/10.1093/advances/nmaa094.
[27] Cui J, Zhou B, Ross SA, Zempleni J. Nutrition, microRNAs, and human health. Adv Nutr (Bethesda, MD) 2017;8(1):105–12. Available from: https://doi.org/10.3945/an.116.013839.
[28] Zempleni J, Aguilar-Lozano A, Sadri M, et al. Biological activities of extracellular vesicles and their cargos from bovine and human milk in humans and implications for infants. J Nutr 2017;147(1):3–10. Available from: https://doi.org/10.3945/jn.116.238949.

[29] Liao Y, Du X, Li J, Lönnerdal B. Human milk exosomes and their microRNAs survive digestion in vitro and are taken up by human intestinal cells. Mol Nutr Food Res 2017;61(11). Available from: https://doi.org/10.1002/mnfr.201700082.
[30] Kosaka N, Izumi H, Sekine K, Ochiya T. microRNA as a new immune-regulatory agent in breast milk. Silence 2010;1(1):7. Available from: https://doi.org/10.1186/1758-907x-1-7.
[31] Chen X, Gao C, Li H, et al. Identification and characterization of microRNAs in raw milk during different periods of lactation, commercial fluid, and powdered milk products. Cell Res 2010;20(10):1128–37. Available from: https://doi.org/10.1038/cr.2010.80.
[32] Benmoussa A, Ly S, Shan ST, et al. A subset of extracellular vesicles carries the bulk of microRNAs in commercial dairy cow's milk. J Extracell Vesicles 2017;6(1):1401897. Available from: https://doi.org/10.1080/20013078.2017.1401897.
[33] Maburutse BE, Park MR, Oh S, Kim Y. Evaluation and characterization of milk-derived microvescicle isolated from bovine colostrum. Korean J Food Sci Anim Resour 2017;37(5):654–62. Available from: https://doi.org/10.5851/kosfa.2017.37.5.654.
[34] Liu X, Zhan Z, Xu L, et al. MicroRNA-148/152 impair innate response and antigen presentation of TLR-triggered dendritic cells by targeting CaMKIIα. J Immunol 2010;185(12):7244–51. Available from: https://doi.org/10.4049/jimmunol.1001573.
[35] Banerjee S, Xie N, Cui H, et al. MicroRNA let-7c regulates macrophage polarization. J Immunol 2013;190(12):6542–9. Available from: https://doi.org/10.4049/jimmunol.1202496.
[36] Lönnerdal B. Human milk microRNAs/exosomes: composition and biological effects. Nestle Nutr Inst Workshop Ser 2019;90:83–92. Available from: https://doi.org/10.1159/000490297.
[37] Donovan SM. Role of human milk components in gastrointestinal development: current knowledge and future needs. J Pediatr. 2006;149(5, Supplement):S49–61. Available from: https://doi.org/10.1016/j.jpeds.2006.06.052.
[38] Ballard O, Morrow AL. Hum milk composition: nutrients and bioactive factors. Pediatr Clin North Am 2013;60(1):49–74. Available from: https://doi.org/10.1016/j.pcl.2012.10.002.
[39] Kim SY, Yi DY. Components of human breast milk: from macronutrient to microbiome and microRNA. Clin Exp Pediatr 2020;63(8):301–9. Available from: https://doi.org/10.3345/cep.2020.00059.
[40] Dvorak B, Fituch CC, Williams CS, Hurst NM, Schanler RJ. Increased epidermal growth factor levels in human milk of mothers with extremely premature infants. Pediatr Res 2003;54(1):15–19. Available from: https://doi.org/10.1203/01.Pdr.0000065729.74325.71.
[41] Philipps AF, Dvorák B, Kling PJ, Grille JG, Koldovský O. Absorption of milk-borne insulin-like growth factor-I into portal blood of suckling rats. J Pediatr Gastroenterol Nutr 2000;31(2):128–35. Available from: https://doi.org/10.1097/00005176-200008000-00008.
[42] Kling PJ, Taing KM, Dvorak B, Woodward SS, Philipps AF. Insulin-like growth factor-I stimulates erythropoiesis when administered enterally. Growth Factors 2006;24(3):218–23. Available from: https://doi.org/10.1080/08977190600783162.
[43] Rodrigues DM, Li AY, Nair DG, Blennerhassett MG. Glial cell line-derived neurotrophic factor is a key neurotrophin in the postnatal enteric nervous system. Neurogastroenterol Motil 2011;23(2):e44–56. Available from: https://doi.org/10.1111/j.1365-2982.2010.01626.x.
[44] Li R, Xia W, Zhang Z, Wu K. S100B protein, brain-derived neurotrophic factor, and glial cell line-derived neurotrophic factor in human milk. PLoS One 2011;6(6):e21663. Available from: https://doi.org/10.1371/journal.pone.0021663.

[45] DiBiasie A. Evidence-based review of retinopathy of prematurity prevention in VLBW and ELBW infants. Neonatal Netw 2006;25(6):393–403. Available from: https://doi.org/10.1891/0730-0832.25.6.393.
[46] Soubasi V, Kremenopoulos G, Diamanti E, Tsantali C, Sarafidis K, Tsakiris D. Follow-up of very low birth weight infants after erythropoietin treatment to prevent anemia of prematurity. J Pediatr 1995;127(2):291–7. Available from: https://doi.org/10.1016/s0022-3476(95)70313-6.

CHAPTER FOUR

Immunoglobulins and their relevance in human milk

"Let not thy food be thy medicine and the medicine the thy food"

Hippocrates

"Look at any country that's plagued with poverty, disease, or violence; the antidote is girls. Girls are the antibodies to many of society's ills"

Queen Rania of Jordan

Key Points

1. Antibody composition differs between various mammalian species.
2. Antibodies in human milk perform their function primarily at the intestinal mucosal surface.
3. There is a critical interaction between intestinal microbes and the antibodies provided by human milk that provides specialized protection for the infant against infectious diseases.

Introduction

Antibodies are protein molecules produced by plasma cells (sometimes referred to as B cells) that are critical components of the immune system. These attach to and inactivate pathogenic microorganisms. Various antibodies are found in human milk. They are essential for the protection of the breastfed infant against infections. There are differences between what is termed an "antibody" and "immunoglobulin." An immunoglobulin has a transmembrane domain attached to the plasma membrane of a B-cell, whereas antibody does not have a transmembrane

The Science of Human Milk.
DOI: https://doi.org/10.1016/B978-0-323-99913-7.00004-6

domain and is found primarily in the circulation. Immunoglobulins are also called surface immunoglobulins. For purposes of this chapter, we will use these terms interchangeably.

Neonatal immunity

Neonates are born with an immature immune system. Transfer of immunity from their mothers provides immune protection. Antibodies are transferred from the mother either via the placenta prenatally or from breast milk postnatally. Placental immunoglobulin G (IgG) is the major antibody transferred via the bloodstream from the mother to the fetus, and it provides its protective effects prenatally. IgG is also found in breast milk for several months postnatally, but wanes over time [1].

After birth, infants are exposed to numerous microbes and environmental antigens. Because of their immature immune systems, neonates depend heavily on sources of immunity derived from its mother. These immune factors also need to distinguish commensal microbes from pathogens.

Evidence supports that antibody transferred via breast milk provides protection against neonatal disease, largely at the intestinal mucosal surface, by preventing infection, selecting the types of commensal microbes present in the intestine, and tolerizing immunity during infancy [1]. Here we will discuss the role of these antibodies in neonatal immunity and development.

Antibody structure

A typical antibody structure is shown in Fig. 4.1. An antibody is a Y-shaped glycoprotein containing several different portions, which have different functions. Two major roles of immunoglobulins are antigen binding and effector function. These functions are located primarily on the variable component of the immunoglobulin. This component binds to specific antigen determinants on the surface of microbial pathogens or pathogen-infected cells in the body.

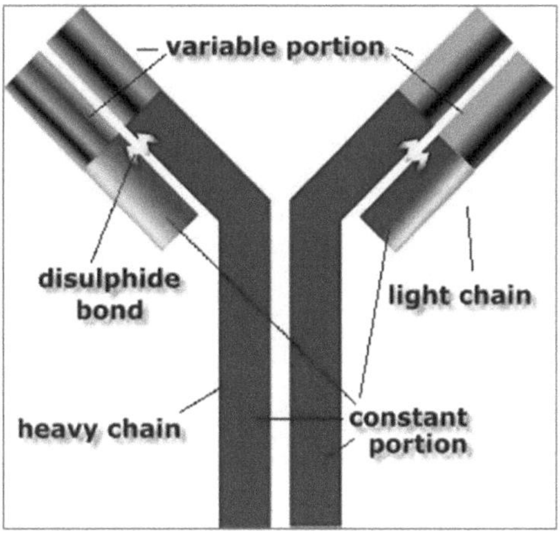

Figure 4.1 Antibody components. *From https://commons.wikimedia.org/w/index.php?curid = 916435.*

The binding of antibodies to a pathogen usually neutralizes the pathogen. Antibodies exhibit two effector functions: (1) They fix a molecule called complement to an antigen–antibody complex, which helps bring about lysis of foreign cells and (2) they bind to various cell types. In this manner, the antibody-bound pathogens along with complement have the ability to trigger macrophages, mast cells, and lymphocytes to induce an immune response against them.

Types of antibodies in milk

Antibodies found in human milk have similarities to those found in other mammalian species. However, the concentrations differ markedly (Fig. 4.2), and an understanding of these differences is critical in appreciating their importance in the human infant.

The production of certain immunoglobulins in the human infant during different developmental stages differs. Because of this, the balance between those derived from the mother and those produced by the infant is of considerable importance. This is particularly important for IgA, which is an important way that the mother confers passive immunity to the neonate. IgA comprises over 90% of all antibodies present in the milk [2].

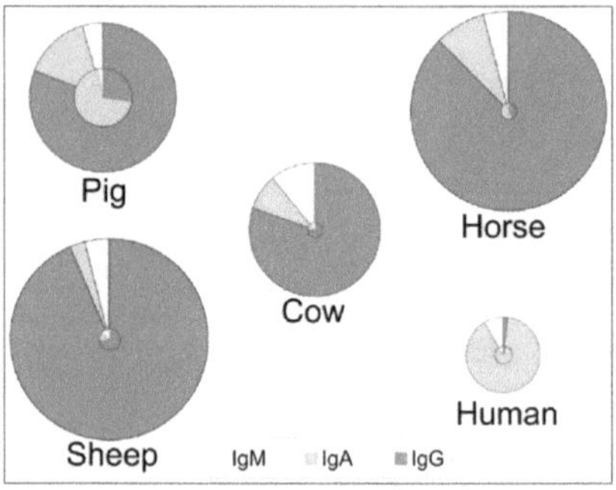

Figure 4.2 Antibody types found in different species.

Although secretory IgA (SIgA) accounts for the majority of breast milk antibodies, IgM and IgG classes are also transferred to the human infant [1]. This difference in immunoglobulin composition across species reflects differences in antibody absorption. Ungulate and rodent infants continue to transfer antibodies from the gut to the circulation for up to 48 h and 21 days after birth, respectively [3]. However, in humans, the intestinal epithelial surface no longer absorbs macromolecules at the time of birth in term-born humans and prevents large macromolecules from passing through [4]. Preterm infants are an exception, where gut closure occurs later after birth [5]. This allows IgG from the breast milk to reach circulation in the first days of life in preterms. In term infants, due to the inability of antibodies in human breast milk to reach circulation, the primary role of these immunoglobulins is to provide intestinal mucosal barrier immunity [6].

A special form of IgA called "secretory" IgA

Human milk contains immunoglobulins of the class M, G, D, and A. Concentrations of IgM, IG, and ID are relatively low in milk compared to serum [7]. IgE is undetectable. SIgA comprises more than 95% of human milk immunoglobulins [8]. These are primarily directed against foreign enteric and respiratory microbial antigens. The quantity of SIgA in human milk declines as lactation proceeds, and the concentrations of SIgA are highest in colostrum and gradually plateau later in lactation. The approximate mean daily intake of SIgA in healthy-term breastfed infants is about ½ gram per day between 1 and 4 months [8].

Antibody function

After birth, an infant is exposed to an array of microbes and environmental antigens that it was not exposed to prenatally. At the time of birth, the neonate's immune system is still immature and relies on sources of immunity gained from the mother.

While in utero, maternal immunity is transferred via the placenta largely as IgG. This passes directly into the serum of infants [7]. This is known to be protective against many childhood diseases and is instrumental in protecting against many childhood infections [9].

Antibodies transferred through the placenta diminish over the first year after birth. Infants who are breastfed continue to receive a supply of maternal immunity through the milk. The immunity supplied by breast milk is largely different than that supplied by the placenta in that rather than providing systemic immunity provided via the placenta, breast milk antibodies confer mucosal immunity. The mucosal barriers close at birth in humans and this prevents breast milk antibodies from transferring into the bloodstream providing systemic immunity. Instead, these antibodies populate the mucosal surfaces of the neonate and enhance the mucosal first line of defense, where they decrease adverse infant outcomes including diarrheal disease, respiratory infections, asthma, diabetes, and inflammatory bowel disease.

Because of the health benefits infants receive through breastfeeding, especially those related to immune protection, exclusive breastfeeding for the first 6 months after birth is recommended by several expert agencies including the World Health Organization.

Our understanding of the role of antibodies transferred from the mother to her infant during breastfeeding is still not clearly understood, but emerging data suggest enrichment of certain populations of antibodies that may have specific roles against specific antigens the mother may have had previous exposure to.

Origins and transfer of antibodies in breast milk

Breast milk antibodies evolve throughout lactation in response to the needs of the infant. Milk produced in the first few days after birth, colostrum, is rich in proteins and immunoglobulins, along with several other immune factors, but is relatively low in lactose. After the colostral phase, transitional milk is formed, which is a rich food source that is highly suitable for meeting the nutritional needs of the rapidly growing infant [10]. The milk continues to mature after a few weeks and contains hormones, growth factors, enzymes, and numerous other bioactive components discussed in other chapters in this book.

Compared to near immediate closure to the passage of macromolecules from milk in most mammalian species, gut closure occurs later after birth in preterm primate infants, permitting IgG from the breast milk to reach circulation in the first days after delivery.

In humans, immune transfer consists of an in utero phase, where antibodies cross the placenta into the fetal bloodstream, and a postbirth phase, where milk antibodies provide mucosal (barrier) immunity to the human infant [6].

IgAs in breast milk are produced by IgA plasma cells that accumulate in the lactating mammary glands. These plasma cells are primed in the lymph nodes and the Peyer's patches of the mucosal tissues via exposures to various antigens, then home to the mammary glands due to expression of the mucosal chemokine (a plasma cell chemoattractant) in the lactating mammary gland, which is upregulated by the mammary gland epithelial cells during lactation [11].

This breast milk IgA provides neonates with similar mucosal barrier activity as found in the mother. These transferred antibodies can provide a barrier against the antigens encountered in her environment.

Antibodies are transferred from the plasma cells in the bloodstream to the breast milk through mammary glands that consist of lactiferous ducts and alveoli, which are involved in delivering milk to the newborn. In this transfer process, plasma cells from mucosal tissues home to the mammary gland. They produce IgA (SIgA) and IgM, which are passed into breast milk by the polymeric Ig receptor (pIgR) on the breast epithelium. IgG originating from the blood is transferred into the breast milk via FcRn expressed on the mammary epithelium cells.

Ductal and alveoli epithelial cells express the pIgR on their basolateral side and express neonatal Fc receptor (FcRn) endosomally [12]. PIgR appears to be a carrier for the passage of SIgA and SIgM into the breast milk. This is accomplished by pIgR binding to polymeric IgA or IgM. The pIgR–antibody complex then undergoes endocytosis (being engulfed into the cell), undergoes transport to the apical surface, and the pIgR is cleaved but a small part of it is still bound to the antibody, which has been called the secretory component. This secretory component helps protect secretory antibodies from cleavage by enzymes in the intestinal tract [13].

Secretory IgA in breast milk

IgA in the mucosal surface is a polymer. Serum IgA is a monomer. SIgA is one of the polymeric IgAs composed of dimeric IgA, J chain, and secretory component (see Fig. 4.3). Most of the sIgAs are generated by the intestinal tract and effects are localized to the intestinal mucosal surface. IgA is thought as a noninflammatory antibody at these mucosal sites. Its polymeric structure contains oligosaccharide side chains of the secretory component which contribute to this property. sIgA is concentrated in the mucosal layer where it crosslinks microorganisms and causes them to clump together. sIgA has the capability to entrap microorganisms. Peristaltic bowel movements or respiratory movements (coughing and sneezing) help remove the bacteria clumps. The sum of processes called immune exclusion is comprised of agglutination, entrapment, and clearance processes.

All newborns lack sIgA at birth, which causes a reliance on the mothers' milk for sIgA. Deficiencies of sIgA are associated with several immune deficiencies and are thought to play a critical role in the

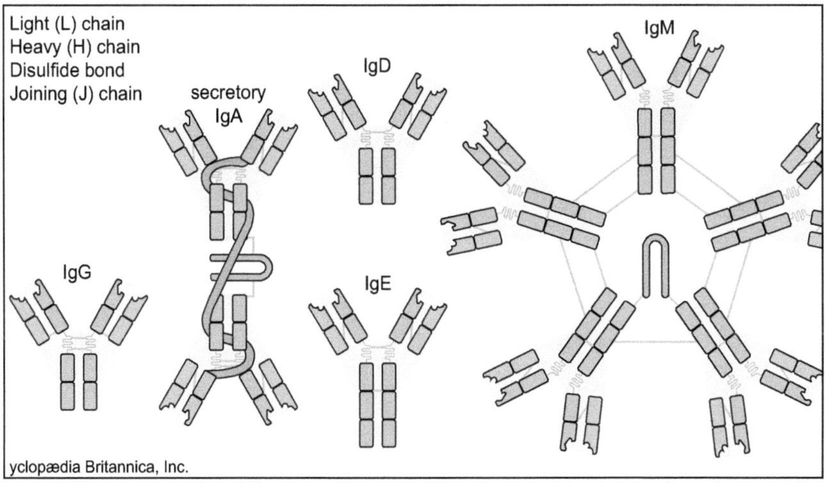

Figure 4.3 The five antibody classes are IgG, IgM, IgA, IgD, and IgE. *From https://www.britannica.com/science/immune-system/Classes-of-immunoglobulins.*

protection of infants born prematurely [14]. In the first several months after birth, the baby's immune system begins to make more IgA, but during the first several weeks, the baby receives IgA from breast milk.

The importance of IgA lies in its capability to coat and seal the infant's respiratory and intestinal tracts. This thwarts the entry of microbes, including bacteria, viruses, fungi, and parasites, into the bloodstream.

Breast milk antibodies and the infant microbiome

The developing microbiome has been shown to have a major impact on long-term health and development [1]. Associations have been found between various aspects of the developing microbiome during infancy and subsequent autoimmune disease, asthma, allergies, and diarrheal illnesses [1]. In utero, infants are clearly exposed to components of the mother's microbiome. During and shortly after birth, the infant is exposed to a massive influx of microorganisms from the mother's vaginal canal, skin-derived nursing microbiome, breast milk, and the environment.

During early infancy, a newborn's microbiome has a low diversity but is susceptible to the introduction of new species. Numerous factors can significantly impact the compositional development of the neonatal microbiome. These include mode of delivery (C-section vs vaginal), gestational age, maternal diet, maternal or neonatal use of antibiotics, and formula versus breastfeeding. Dysbiosis and/or the effects of improper seeding may persist throughout life. One example of altered seeding that can be long-lasting is related to the mode of delivery. Infants born by C-section have an increased risk of asthma allergy, type I diabetes, and obesity, all associated with an imbalanced microbiome [15].

SigA and the microbiome

SIgA plays a critical role in the establishment and maintenance of the microbiome during infancy. Studies suggest that SIgA can bind antigens through glycans on the Fc portion of the antibody and on the SC and J-chain, allowing for broad reactivity of SIgA to antigens. SIgA traps microbes and prevents the translocation of both commensal and pathologic microbes through the mucosal epithelia by slowing attachment and separating microbiota from the epithelium. This process is called immune exclusion.

Breast milk antibodies recognize and segregate commensals from the gut epithelium.

Penetration of bacteria through the epithelium is prevented by these maternal antibodies, which limits an immune response to these microbes. It also provides signals for underlying antigen-presenting cells to sample complexes of microbes targeted for tolerance. When maternal antibodies are absent, bacteria can penetrate the neonatal gut epithelium. This results in activation of T cells in the mesenteric lymph nodes underlying the epithelium, with the subsequent induction of a T cell-dependent antibody response.

IgA deficiency has been associated with elevated rates of allergies and autoimmune diseases. This may be related to the altered microbiota in these infants [14]. SIgA not only protects against pathogens in the gut but also drives the development and maintenance of a healthy gut microbiota.

Microbiome colonization and breast milk IgA

Production of neonatal-derived SIgA begins several months after birth and does not reach adult levels until 3 years of infant age. Maternal SIgA from BM is the only source of SIgA during early life, bridging the time until the infant produces endogenous SIgA, and represents a regulator of infant microbial colonization. Breastfed infants have a different microbiome composition compared to formula-fed infants [16].

Antibodies in milk and infections

Antibodies are a known benefit of breastfeeding. When breast milk coats the baby's oral mucosa, nasal cavity, Eustachian tubes, and GI tract, the IgA binds to bacteria and viruses at that surface immobilizing the invasive organism. These antibodies are produced specifically to recognize and attack antigens (parts of the foreign pathogen[s]) currently present in your environment. Some antibodies can remember the antigen and facilitate a quicker immune response the next time you are exposed to that pathogen. This is a component of our immune system called the adaptive immune response.

There are different types of antibodies found in milk, which include IgG, IgM, and IgA. Not all these antibodies are the same. During infection, the first antibodies produced are IgM antibodies and are transiently found in the blood and lymph. Both IgM, and later IgG, initiate pathways that mount a systemic response. IgM is the fastest antibody produced, but it is short-lived. IgG antibodies are the most common antibody and are the only type of antibody that can cross the placenta of a pregnant woman in order to help protect the fetus.

IgA is stabilized with a secretory component when it is transported across mucous membranes, becoming SIgA. The overall job of specific SIgA is to inhibit initial pathogen colonization by performing immune exclusion both on the mucosal surface and within virus-infected secretory epithelial cells, all without causing tissue damage. In short, SIgA antibodies function by binding to pathogens that come in contact throughout the GI tract, effectively neutralizing them. Binding these foreign substances prevents them from entering the body. In essence, you can think of SIgA antibodies as our "first line of defense." SIgA antibodies appear to be highly resistant to digestion by enzymes that are in the GI tract.

IgA is secreted into the blood, GI tract, and mucous membranes of mothers and infants. Most antibodies are too large to transfer into breast milk in large amounts, but IgA is capable of being transported across the lactocyte barrier into the milk.

Maternal IgA consumed from a breastfed infant can help protect an infant as they build their own immune defenses. A mother and her child are often in the same environments and therefore exposed to the same pathogens. This is a huge benefit for antibodies in breast milk because they are customized to protect the baby against the specific pathogens they are exposed to, called the enteromammary system.

Human milk is dynamic in its response to viruses

Several studies have suggested the existence of an enteromammary system [17,18] (Fig. 4.4), which relates the changes in the maternal immune system to colonization or infection in the mother or infant by microorganisms, followed by recognition of the antigens by the mother's immune system, a response by the maternal immune system and a subsequent passage of relevant immune agents to the infant through the breast milk.

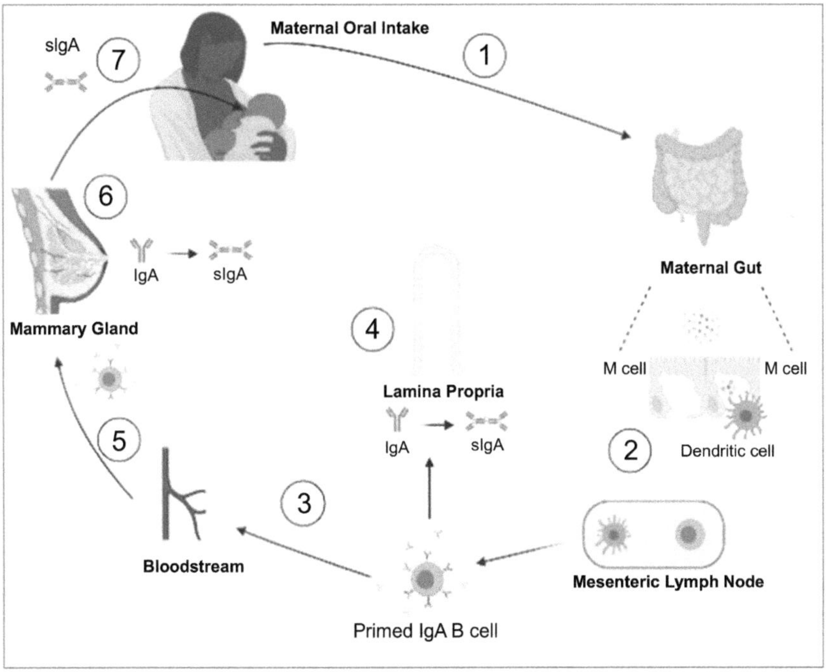

Figure 4.4 The enteromammary system. *Adapted from Newburg DS, Allan Walker W. Protection of the neonate by the innate immune system of developing gut and of human milk. Pediatr Res 2007;61(1):2–8.*

As seen in this figure, the enteromammary circulation allows the transportation of maternal immunoglobulins from the mother via her milk into the infant. Immune cells primed by viruses and other potentially pathogenic agents produce sIgA, which in turn provides protection to the infant's intestinal tract. The IgAs are transported across alveolar epithelial cells by pIgRs. There, they bind to the secretory components to form sIgAs; (**7**) sIgA passes into the infant gut lumen to provide protection from microbes through a range of mechanisms [17].

Immunoglobins in breast milk related to COVID-19 infections and vaccinations

Pregnancy has been considered as a risk factor for the severity of disease with COVID-19. SARS-CoV-2 intrauterine transmission seems to be rare. However, most studies show COVID-19 during pregnancy increases the risk of pregnancy complications [19]. COVID-19 vaccination during pregnancy is safe and effective [19]. Antibodies to SARS-CoV-2 have been found in breast milk that shows neutralizing capability when extracted from vaccinated infants' stools following maternal vaccination [20]. These studies also measured and tested antibodies found in the mothers' blood plasma and breast milk soon after vaccination and then again about 6 months later. The researchers found that the antibodies in the plasma and milk of vaccinated people were better able to neutralize the virus, though they also observed that antibody levels decreased at the 6-month mark, which other vaccine studies have found as well. It is highly likely that this provides protection to the infant [21].

Conclusions

Maternal antibody transferred via breast milk confers protection against pathogens and plays an important role in shaping life-long immunity.

IgA plays a central role in regulating the microbiome. There is also a major role for IgG in fighting pathogens. Furthermore, subpopulations of breast milk-derived IgA, IgG, and other iso-types and subclasses also shape neonatal immunity. The specific mechanisms of these interactions remain to be further elucidated.

References

[1] Atyeo C, Alter G. The multifaceted roles of breast milk antibodies. Cell 2021; 184(6):1486–99. Available from: https://doi.org/10.1016/j.cell.2021.02.031.
[2] McElroy SJ, Weitkamp JH. Innate immunity in the small intestine of the preterm infant. Neoreviews 2011;12(9):e517–26. Available from: https://doi.org/10.1542/neo.12-9-e517.
[3] Sangild PT, Trahair JF, Loftager MK, Fowden AL. Intestinal macromolecule absorption in the fetal pig after infusion of colostrum in utero. Pediatr Res 1999; 45(4):595–602.
[4] Van de Perre P. Transfer of antibody via mother's milk. Vaccine. 2003;21(24):3374–6.
[5] Weaver L, Laker M, Nelson R. Intestinal permeability in the newborn. Arch Dis Child 1984;59(3):236–41.
[6] Brandtzaeg P. Induction of secretory immunity and memory at mucosal surfaces. Vaccine 2007;25(30):5467–84. Available from: https://doi.org/10.1016/j.vaccine.2006.12.001.
[7] Ogra PL. Immunology of human milk and lactation: historical overview. Nestle Nutr Inst Workshop Ser 2020;94:11–26. Available from: https://doi.org/10.1159/000505211.

[8] Palmeira P, Carneiro-Sampaio M. Immunology of breast milk. Rev Assoc Med Bras 1992;62(6):584–93. Available from: https://doi.org/10.1590/1806-9282.62.06.584.
[9] Healy CM. Vaccines in pregnant women and research initiatives. Clin Obstet Gynecol 2012;55(2):474–86. Available from: https://doi.org/10.1097/GRF.0b013e31824f3acb.
[10] Lyons KE, Ryan CA, Dempsey EM, Ross RP, Stanton C. Breast milk, a source of beneficial microbes and associated benefits for infant health. Nutrients 2020;12(4):1039.
[11] Wilson E, Butcher EC. CCL28 controls immunoglobulin (Ig)A plasma cell accumulation in the lactating mammary gland and IgA antibody transfer to the neonate. J Exp Med 2004;200(6):805–9. Available from: https://doi.org/10.1084/jem.20041069.
[12] De Groot N, Van Kuik-Romeijn P, Lee SH, De Boer HA. Increased immunoglobulin A levels in milk by over-expressing the murine polymeric immunoglobulin receptor gene in the mammary gland epithelial cells of transgenic mice. Immunology 2000;101 (2):218–24. Available from: https://doi.org/10.1046/j.1365-2567.2000. 00094.x.
[13] Demers-Mathieu V, Nielsen SD, Underwood MA, Borghese R, Dallas DC. Changes in proteases, antiproteases, and bioactive proteins from mother's breast milk to the premature infant stomach. J Pediatr Gastroenterol Nutr 2018;66(2):318–24. Available from: https://doi.org/10.1097/mpg.0000000000001719.
[14] Schaffer FM, Monteiro RC, Volanakis JE, Cooper MD. IgA deficiency. Immunodefic Rev 1991;3(1):15–44.
[15] Rushing J, Neu J. Cesarean versus vaginal delivery: long-term infant outcomes and the hygiene hypothesis. Clin Perinatol 2011;38(2):321–31.
[16] Davis EC, Wang M, Donovan SM. The role of early life nutrition in the establishment of gastrointestinal microbial composition and function. Gut Microbes 2017;8 (2):143–71. Available from: https://doi.org/10.1080/19490976.2016.1278104.
[17] Newburg DS, Allan Walker W. Protection of the neonate by the innate immune system of developing gut and of human milk. Pediatr Res 2007;61(1):2–8.
[18] Kleinman RE, Walker WA. The enteromammary immune system: an important new concept in breast milk host defense. Dig Dis Sci 1979;24(11):876–82.
[19] Rasmussen SA, Jamieson DJ. COVID-19 and pregnancy. Infect Dis Clin North Am 2022;36(2):423–33. Available from: https://doi.org/10.1016/j.idc.2022.01.002.
[20] Stafford LS, Valcarce V, Henry M, et al. Detection of SARS-CoV-2 IgA and IgG in human milk and breastfeeding infant stool 6 months after maternal COVID-19 vaccination. J Perinatol 2023;43:775–81. Available from: https://doi.org/10.1038/s41372-022-01581-5.
[21] Valcarce V, Stafford LS, Neu J, et al. Detection of SARS-CoV-2-Specific IgA in the human milk of COVID-19 vaccinated lactating health care workers. Breastfeed Med 2021;16(12):1004–9. Available from: https://doi.org/10.1089/bfm.2021.0122.

CHAPTER FIVE

Multiomics of human milk

Key points

1. Individual "omics" studies related to human milk are of interest and provide knowledge about associations between each and the phenotype of the individual infant. However, associations do not provide clear mechanistic information about causal pathways. Integrating several layers of omics provides the opportunity to better determine how these interact and allow for a better understanding of their mechanisms.
2. The dynamic nature and emergent properties of the different "omics" highlight the necessity of longitudinal multilayer studies using tailored computational methods, which will eventually allow us to identify and characterize milk interactions and dynamics that are associated with optimal health outcomes.

Introduction

Other chapters in this book discuss various bioactive components of human milk that include the immunoglobulins, bioactive proteins, and oligosaccharides. In this chapter, we discuss a relatively new area related to human milk composition, the microbiome, metabolome as well as other "omic" components and their functions.

The cartoon at the beginning of the chapter depicts a parable of six blind men, each of whom was asked to identify a certain structure. One felt the side and stated, "it's a wall," another felt a tail and stated "it's a rope," another felt the tusk and stated "it's a spear," and so on. They were each very dogmatic in stating what they felt, but they were not able to identify that the structure they felt was an elephant. Had they collaborated and synthesized their ideas, they may have been able to figure out this was an elephant. Similarly, recent research has provided us with outstanding

The Science of Human Milk.
DOI: https://doi.org/10.1016/B978-0-323-99913-7.00010-1

new tools to evaluate the human genome, microbial DNA and RNA sequences, metabolites, immunologic components, epigenetic modifications, and various other "omics." Evaluation of each one of these omics individually, as with the assessment of each of the blind men, can provide valuable information about associations, but by itself is not able to determine the entire structure or mechanistic process. Working together in a multiomic (or collaborative blind men) approach is likely to provide valuable information regarding mechanisms, well beyond associative science.

In this chapter, we will provide information regarding several of these omic approaches in the evaluation of human milk. We will not get into details about the technical analyses but rather attempt to underline the fact that integrating these omics will be an important feature in understanding how the compositional and dynamic changes in milk might affect the infant as well as providing biomarkers for precision nutrition and aiding the industry in improving their products for when mother's milk may not be available.

The microbiome

Human milk has traditionally been considered a sterile fluid and prior to the last decade, it was considered highly problematic if microbes were found in human milk as they were considered a consequence of poor hygiene and possible carriers of disease [1]. In 2003 the finding of endogenous lactic acid bacteria from eight healthy mothers suggested that these microbes may be symbionts that provide a role in the prevention of neonatal infectious diseases [2].

More recently, the use of culture-independent techniques that do not rely on taxa-specific growth media but instead identify bacterial groups because of variation in specific DNA segments has led to a scientific paradigm shift. These studies using molecular techniques overwhelmingly support the existence of a rich and diverse community of bacteria in milk [3].

We are beginning to recognize human milk as a live ecosystem containing bacteria, viruses, and fungi, together known as the microbiome. How these microbes are seeded into the milk and their function for the infant and the lactating mother are under intense investigation. Numerous studies support a strong interaction of the milk microbes, the developing immune system, and various metabolic and epigenetic mechanisms. Exactly how these interactions shape future health remains a topic of intense investigation. It is clearly important that we develop a better understanding of this human milk microbiome and interaction with other omics, the host, and the environmental exposures that may modify them.

Here we will first discuss an overview of the human milk microbiome, where the microbes come from, what they do, and what they produce in terms of metabolites and how they interact with the developing immune system. We will also discuss their potential role in future health and disease.

What bacteria are found in human milk?

Perhaps the first of the studies that used culture-independent methods to study healthy human milk microbiota was published in 2003 by Dr. Juan Rodríguez and his research group [2]. This study tested the hypothesis that human milk is a source of lactic acid bacteria. It combined targeted culture-dependent and culture-independent methods. Lactic acid bacterium types in the milk differed from those found on the skin of the breast. This led to the conclusion that human milk bacteria have an endogenous origin rather than being a contaminant from the skin.

Additional studies provide indisputable evidence that human milk contains a diverse and viable bacterial community [3–5]. Most of the studies have shown that milk contains *Staphylococcus*, *Streptococcus*, and *Lactobacillus* genera. Whether differences in bacterial community structure in human milk are due to environmental, behavioral, or genetic differences or are a consequence of methodologic variation requires additional study.

Where do they come from?

There are three hypothesized sources of the human milk microbiome [6]. One potential origin is the infant's oral cavity. Microbes such as streptococci derive from retrograde flow of milk from the infant oral cavity to mammary ducts. A second putative mechanism is that they originate from the human skin. Skin microbes such as Staphylococcus epidermidis may colonize the mammary gland by maternal skin microbiota through the nipples. A third potential source is the maternal gastrointestinal tract from which largely anaerobic bacteria such as Bifidobacteria, Bacteroides, and Clostridia are derived. These are thought to be derived from dendritic cells from the gastrointestinal tract during the latter stages of pregnancy wherein live bacteria are transported through the lymphatic circulation. This is sometimes referred to as the enteromammary pathway (discussed in Chapter 4). Currently, the retrograde flow and the enteromammary pathway are accepted as the two primary sources of origin of milk microbiota [7]. Retrograde flow results from transmission of microbes from the infant's oral cavity into the mammary duct during suckling. The enteromammary derivation is from the translocation of maternal gut bacteria through the intestinal epithelial barrier to reach the mammary gland via the lymphatic circulation.

Types of microbes

Bacteria, fungi, and viruses are found in human milk. Using newly developed technologies that involve gene sequencing of microbes, there are thought to be many different species of bacteria and milk. Of major interest, there is large interindividual variability as well as variability that is related to geographic location, and nuances of the methodologies used to evaluate the milk microbes [8–10].

Staphylococcus, *Streptococcus*, *Serratia*, *Pseudomonas*, *Corynebacterium*, *Ralstonia*, *Propionibacterium*, *Sphingomonas*, and *Bradyrhizobium* are the nine genera that constitute the "core" bacteriome of the human milk microbiota (HMM). They represent approximately half of the microbial milk community [11,12]. The mammary glands, periareolar skin, and infant's mouth may influence its composition [13]. Every day, a breastfed infant ingests between 1×10^5 and 1×10^7 bacteria, which means that almost 30% of the infant's bacteria are acquired through breast milk [5].

When using and interpreting DNA-based technologies, it is important to note that they do not differentiate between live or dead bacteria. One needs to evaluate the viability of bacteria using other technologies such as culture, which demonstrates their capability for reproduction and metabolism. However, the exact culture media and other environmental components required for some of these microbes may not be readily known.

There are also many viruses found in human milk the majority of which are bacteriophages [14]. Their pattern can change in relation to the mode of delivery, birth weight, and lactational stage. A significant number of shared viruses have been identified between milk and infant stool from mother–infant pairs which supports the fact that these infant stool microbes are likely derived from maternal milk [15]. Fungi are also found in milk. The function of these is poorly understood [16].

Factors that influence the composition of the milk microbiome

Numerous environmental factors can affect the composition of the human milk microbiome [12]. These include maternal exposures such as antibiotic use, stress, mode of delivery, diet, geographic location, the season of the year, methods used to collect the milk, stage of lactation, heat, or other treatments such as pasteurization and ultraviolet exposure, and gestational age. Even the gender of the infant may play a role [17] (Fig. 5.1).

Here we discuss some of these factors that shape the HMM

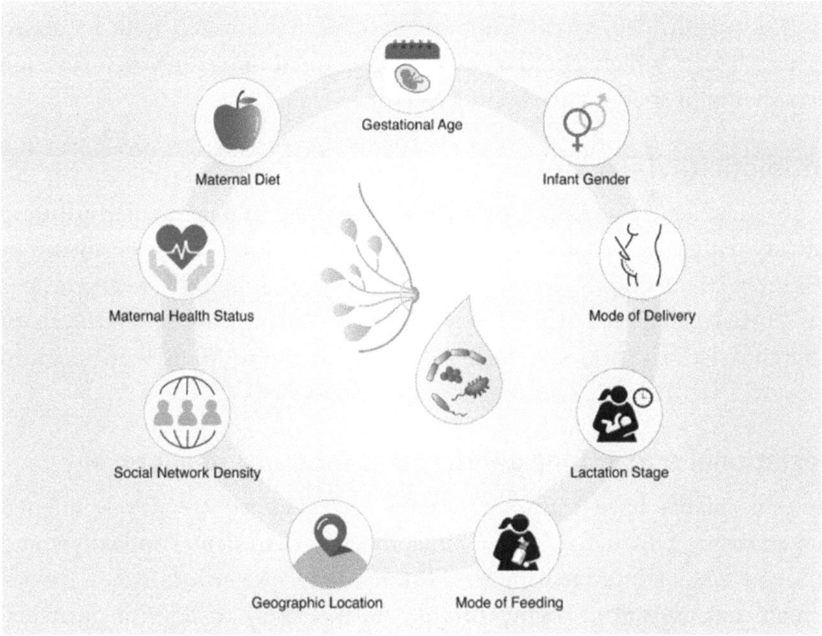

Figure 5.1 Factors affecting the microbiota of human milk. *From Mantziari A., Rautava S. Factors influencing the microbial composition of human milk. Semperinatology 2021;45(8):151507.*

Maternal diet

It is well known that maternal diet can influence the bacterial taxonomic composition of human milk. Since many of its nutrients may be utilized by bacteria [18]. There are positive correlations with polyunsaturated fatty acids, certain vitamins, protein consumption, and vitamin intake.

Maternal health status

Maternal obesity is a major risk factor for childhood obesity. Microbial composition between obese and normal-weight mothers may play a role. Overweight and obese mothers produce milk with a lower microbial diversity and lower counts of Bifidobacterium and a higher level of *Staphylococcus, Akkermansia*, and *Granulicatella* than normal-weight women [18]. These changes may affect the human milk metabolome and further influence the infant microbiome development.

Stress

A primary concern for women in the postpartum period is maternal postnatal distress, which is characterized by symptoms of depression or anxiety. Postpartum depressive symptoms have been associated with increased cortisol levels, which could alter the fecal microbiota diversity through the recently proposed gut–brain axis [18].

Antibiotic therapy

Antibiotic therapy is common during pregnancy and particularly during delivery either as prophylaxis to protect the mother and her infant or treatment of infection. The use of antibiotics has significant effects on several HMM parameters. One study described an increase in the levels of different bacterial species and their diversity in the mothers who received antibiotics [18,19].

Gestational age, infant gender, and mode of delivery

Preterm infants have immune systems characterized by developmental immaturities. This makes them susceptible to detrimental infections and various forms of intestinal injuries. Interestingly, preterm infants fed with human milk instead of infant formula are less likely to develop intestinal injuries. A potential mechanism behind this finding is that some beneficial human milk bacteria may colonize the infant gut, assisting in the development of the immune system and thus providing protection against infections. However, significant differences in the HMM according to the gestational age have been reported, with higher levels of *Enterococcus* and lower levels of *Bifidobacterium* observed in the milk of preterm-delivering mothers compared to those who gave birth on the expected birth date.

Mode of delivery is reportedly associated with profound compositional changes of the HMM. Although human milk from C-section mothers is correlated with higher total bacterial counts these are less diverse with lower levels of *Bifidobacterium, Lactobacillus* compared to the milk of mothers who underwent a vaginal delivery [19].

Geographic location and social network density

Various studies have explored the impact of geographic location on the HMM composition. When the HMM composition was studied in mothers who had delivered vaginally, Spanish mothers had the highest relative abundance of Bacteroidetes in milk when compared to mothers from Finland, China, or South Africa. In the same study, Chinese women who delivered by C-section harbored the highest levels of Actinobacteria in their milk compared to their European or African counterparts.

Dynamic changes over time

The three stages of milk development, colostrum, transitional and mature stages, undergo alterations in the microbiome. Different patterns over time have been reported in different studies. Whether macronutrients such as lactose concentration, human milk oligosaccharides (HMOs), milk fatty acids, hormones, antibodies, or other factors affect these changes over time remains poorly understood [17]. Whether infants are delivered term versus preterm plays a role in terms of microbial taxonomic composition. Lower counts bifidobacterium species have been detected in samples for mothers who delivered at term compared to preterm [20].

Maternal factors

Women who undergo cesarian section when compared to vaginal delivery infants show lower bacterial diversity and richness in their milk [21]. They also have lower levels of bifidobacterium and Lactobacillus species in their milk. Antibiotics also affect the microbes in milk with a decrease in the abundance of lactobacilli, bifidobacteria, and Staphylococcus species in those mothers who received perinatal antibiotics [22]. Diet also affects milk composition [23]. Dietary factors include fiber content, fat, and vitamin intake. Prepregnancy body mass index and weight gain during pregnancy also relate to altered abundance of certain bacterial strains in human milk. Postnatal psychological stress also has been linked to lower milk bacterial diversity [24].

Human milk metabolome

The metabolome refers the small molecules found within a biological sample. These small molecules may include metabolites that are produced by the organism, components of foods, molecules produced by microbes, and so on. These include amino acids, organic acids, nucleic acids, fatty acids, amines, sugars, vitamins, cofactors, pigments, antibiotics, drugs, environmental contaminants, food additives, toxins, and other xenobiotics that are not naturally produced by an organism.

Human milk also contains a wide array of metabolites the concentration of which varies and depends on various factors including the microbial composition of the breast milk. A better understanding of the metabolome of breast milk could be crucial in the future programming of metabolic processes in infants, which is crucial in preventing many diseases and maintaining health. The human milk metabolome has been defined as the complete complement of all low-molecular-weight molecules (<1500 Da) within the human milk niche [24]. Some of these are produced by the human host and others by the resident microbes [25]. Here we discuss the nutritional and nonnutritional biologically active metabolites of human milk, and their role in infant immune development, intestinal microbial colonization, and how they relate to health and developmental outcomes. Sources of these molecules may be from microbes, the human host, or interactions between them. Dietary components ingested and metabolized by microbes and/or the host are included in the overall interactive scheme in Fig. 5.2.

This is a relatively new area of investigation. The first metabolomic study on human milk was conducted in 2012 by Marincola et al. [26] using both NMR spectroscopy and gas chromatography-mass spectrometry to study the aqueous and lipid fractions, respectively, of preterm human milk. The extracts of preterm human milk had higher lactose concentrations than the formula, whereas the preterm formula had a higher concentration of maltose.

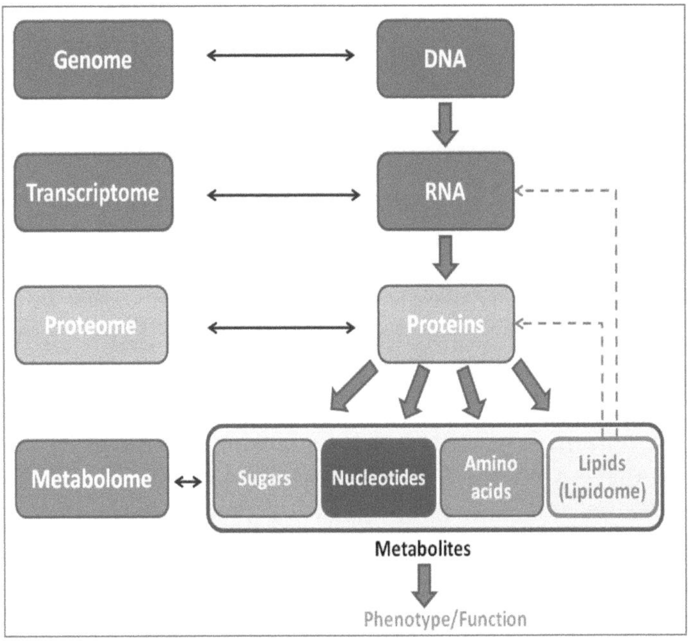

Figure 5.2 The relationships between the genome, transcriptome, proteome, and metabolome (lipidome). *From https://en.wikipedia.org/wiki/File:Metabolomics_schema.png.*

Subsequent studies of metabolites in human milk and formula confirmed a unique metabolic profile of human milk and highlighted the significant differences between human milk and formula milk in relation to both the presence and abundance of most metabolites [27].

Specific metabolites

Carbohydrates

Lactose is the most abundant metabolite and carbohydrate in human milk [28]. As a primary source of calories, lactose provides about 40% of the total energy value of human milk to the infant. Lactose regulates human milk volume because it is the main osmotic component of human milk and draws water into the intracellular secretory vesicles [25]. Lactose concentration ranges between 6.7 and 7.8 g/dL in mature human milk. Its concentration is the least variable of the macronutrients in human milk because it is tightly regulated by the mammary gland [29]. Lactose concentration is correlated with the volume of intake and the number of feeds per day [30,31]. It is also highly related to infant growth [32].

Lactose also forms part of many carbohydrate-based bioactive compounds in human milk (HM), such as forming the backbone of oligosaccharides [33]. Other carbohydrates found in human milk include HMOs. These are described in greater detail in another chapter. A very important aspect of HMOs is that they are complex carbohydrates that are resistant to digestion during their passage through the intestine. Human infants largely lack the enzymes needed to break them down. Certain commensal bacteria in the infant gut are used to digest these complex carbohydrates to produce short-chain fatty acids (SCFAs). These help establish a stable ecosystem in an infant's gut by modulating the immune system and promoting the gut epithelial barrier function [34]. They also drive the development and function of regulatory T (Treg) cells, and thus limit intestinal inflammation. SCFAs, especially butyrate, serve as an energy source for the epithelial cells of the colon and produce an acidic milieu, which makes it inhospitable to potentially pathogenic microbes [35]. SCFAs such as butyrate also regulate gene expression through inhibition of histone deacetylase. This action may suppress nuclear factor kB (NF-kB) activation and thereby leads to the inhibition of interferon-γ production [36]. NF-κB leads to early immune inflammatory responses. Its dysregulation is seen in inflammatory bowel diseases [37].

Lipids

The lipid composition of human milk is highly variable and can be affected by duration of breastfeeding, time of day, stage of lactation, maternal nutritional status, and maternal diet (related to geographic location) [38].

Biogenic amines

Biogenic amines are low-molecular-weight nitrogenous organic bases that are secreted in human milk [39]. They include the polyamines together with the monoamines (tyramine) and diamines. *Enterococcus*, found often in human milk, is the main producers of biogenic amines, mainly putrescine and tyramine. Polyamines inhibit cytokine synthesis in human mononuclear cells, thereby downregulating the proinflammatory cytokine response [40].

Summary and the future

We have just begun recognizing the fact that there is a myriad of microbes in human milk with the potential for exerting numerous beneficial as well as detrimental outcomes for the developing infant. These microbes interact with the human host immune system, produce metabolites, and exert epigenetic effects that may have long-term health outcomes. As mentioned, some of the major questions that need to be answered cannot rely simply on single omic analyses and associating these with phenotypic outcomes. Other omics such as epigenetics, metabolomics, and proteomics will need to be integrated to obtain broader pictures of mechanisms and causality.

Human milk composition represents a complex adaptive system. It is also dynamic and changes partially depending on maternal exposures and other environmental factors. This highlights the necessity of longitudinal multiomic studies using newly developed computational methods. These will allow us to better identify and characterize milk interactions and dynamics associated with optimal health outcomes and provide environments that enhance these.

References

[1] Lyons KE, Ryan CA, Dempsey EM, Ross RP, Stanton C. Breast milk, a source of beneficial microbes and associated benefits for infant health. Nutrients. 2020;12 (4):1039.

[2] Martin R, Langa S, Reviriego C, Jiminez E, Marin ML, Xaus J, et al. Human milk is a source of lactic acid bacteria for the infant gut. J Pediatr 2003;143(6):754–8.

[3] McGuire MK, McGuire MA. Human milk: mother nature's prototypical probiotic food? Adv Nutr (Bethesda, MD) 2015;6(1):112–23.

[4] Fernández L, Pannaraj PS, Rautava S, Rodríguez JM. The microbiota of the human mammary ecosystem. Front Cell Infect Microbiol 2020;10:586667.

[5] Selma-Royo M, Calvo Lerma J, Cortés-Macías E, Collado MC. Human milk microbiome: from actual knowledge to future perspective. Semperinatology 2021;45 (6):151450.

[6] Jeurink P, van Bergenhenegouwen J, Jiménez E, Knippels L, Fernández L, Garssen J, et al. Human milk: a source of more life than we imagine. Benef Microbes 2012;4:17–30.

[7] Lopez Leyva L, Brereton NJB, Koski KG. Emerging frontiers in human milk microbiome research and suggested primers for 16S rRNA gene analysis. Comput Struct Biotechnol J 2021;19:121–33.

[8] McGuire MK, McGuire MA. Got bacteria? The astounding, yet not-so-surprising, microbiome of human milk. Curr Opin Biotechnol 2017;44:63–8.

[9] McGuire MK, McGuire MA. Mapping the human milk microbiome: impetus for a long-awaited renaissance in maternal and infant nutrition research? J Nutr 2020;151 (2):278–80.
[10] Pace RM, Williams JE, Robertson B, Lackey KA, Meehan CL, Price WJ, et al. Variation in human milk composition is related to differences in milk and infant fecal microbial communities. Microorganisms 2021;9(6):1153.
[11] Moubareck CA. Human milk microbiota and oligosaccharides: a glimpse into benefits, diversity, and correlations. Nutrients. 2021;13(4):1123.
[12] Demmelmair H, Jiménez E, Collado MC, Salminen S, McGuire MK. Maternal and perinatal factors associated with the human milk microbiome. Curr Dev Nutr 2020;4 (4):nzaa027.
[13] Groer MW, Morgan KH, Louis-Jacques A, Miller EM. A scoping review of research on the human milk microbiome. J Hum Lactation 2020;36(4):628–43.
[14] Dinleyici M, Pérez-Brocal V, Arslanoglu S, Aydemir O, Sevuk Ozumut S, Tekin N, et al. Human milk virome analysis: changing pattern regarding mode of delivery, birth weight, and lactational stage. Nutrients 2021;13(6):1779.
[15] Pannaraj PS, Ly M, Cerini C, Saavedra M, Aldrovandi GM, Saboory AA, et al. Shared and distinct features of human milk and infant stool viromes. Front Microbiol 2018;9:1162.
[16] Moossavi S, Fehr K, Derakhshani H, Sbihi H, Robertson B, Bode L, et al. Human milk fungi: environmental determinants and inter-kingdom associations with milk bacteria in the CHILD Cohort Study. BMC Microbiol 2020;20(1):146.
[17] Urbaniak C, Angelini M, Gloor GB, Reid G. Human milk microbiota profiles in relation to birthing method, gestation and infant gender. Microbiome 2016;4(1):1.
[18] Mantziari A, Rautava S. Factors influencing the microbial composition of human milk. Semperinatology 2021;45(8):151507.
[19] Soto A, Martín V, Jiménez E, Mader I, Rodríguez JM, Fernández L. Lactobacilli and bifidobacteria in human breast milk: influence of antibiotherapy and other host and clinical factors. J Pediatr Gastroenterol Nutr 2014;59(1):78.
[20] Khodayar-Pardo P, Mira-Pascual L, Collado MC, Martínez-Costa C. Impact of lactation stage, gestational age and mode of delivery on breast milk microbiota. J Perinatol 2014;34(8):599–605.
[21] Cabrera-Rubio R, Mira-Pascual L, Mira A, Collado MC. Impact of mode of delivery on the milk microbiota composition of healthy women. J Dev Origins Health Dis 2016;7(1):54–60.
[22] Rautava S. Early microbial contact, the breast milk microbiome and child health. J Dev Origins Health Dis 2016;7(1):5–14.
[23] Padilha M, Danneskiold-Samsøe NB, Brejnrod A, Hoffmann C, Cabral VP, Iaucci JdM, et al. The human milk microbiota is modulated by maternal diet. Microorganisms. 2019;7(11):502.
[24] Browne PD, Aparicio M, Alba C, Hechler C, Beijers R, Rodríguez JM, et al. Human milk microbiome and maternal postnatal psychosocial distress. Front Microbiol 2019;10:2333.
[25] Smilowitz JT, O'Sullivan A, Barile D, German JB, Lönnerdal B, Slupsky CM. The human milk metabolome reveals diverse oligosaccharide profiles. J Nutr 2013;143 (11):1709–18.
[26] Marincola FC, Noto A, Caboni P, Reali A, Barberini L, Lussu M, et al. A metabolomic study of preterm human and formula milk by high resolution NMR and GC/ MS analysis: preliminary results. J Matern Fetal Neonat Med 2012;25(Suppl 5):62–7.
[27] Qian L, Zhao A, Zhang Y, Chen T, Zeisel SH, Jia W, et al. Metabolomic approaches to explore chemical diversity of human breast-milk, formula milk and bovine milk. Int J Mol Sci 2016;17(12):2128.

[28] Ojo-Okunola A, Cacciatore S, Nicol MP, du Toit E. The determinants of the human milk metabolome and its role in infant health. Metabolites. 2020;10(2):77.
[29] Ballard O, Morrow AL. Human milk composition: nutrients and bioactive factors. Pediatr Clin North Am 2013;60(1):49–74.
[30] Prentice P, Ong KK, Schoemaker MH, van Tol EA, Vervoort J, Hughes IA, et al. Breast milk nutrient content and infancy growth. Acta Paediatr (Oslo, Nor: 1992) 2016;105(6):641–7.
[31] Khan S, Hepworth AR, Prime DK, Lai CT, Trengove NJ, Hartmann PE. Variation in fat, lactose, and protein composition in breast milk over 24 hours: associations with infant feeding patterns. J Hum Lactation 2013;29(1):81–9.
[32] Nilsson A, Mardinoglu A, Nielsen J. Predicting growth of the healthy infant using a genome scale metabolic model. NPJ Syst Biol Appl 2017;3:3.
[33] Martin CR, Ling PR, Blackburn GL. Review of infant feeding: key features of breast milk and infant formula. Nutrients. 2016;8(5):279.
[34] Sun Y, O'Riordan MX. Regulation of bacterial pathogenesis by intestinal short-chain fatty acids. Adv Appl Microbiol 2013;85:93–118.
[35] Fukuda S, Toh H, Hase K, Oshima K, Nakanishi Y, Yoshimura K, et al. Bifidobacteria can protect from enteropathogenic infection through production of acetate. Nature 2011;469(7331):543–7.
[36] Berni Canani R, Di Costanzo M, Leone L. The epigenetic effects of butyrate: potential therapeutic implications for clinical practice. Clin Epigenet 2012;4(1):4.
[37] Canani RB, Costanzo MD, Leone L, Pedata M, Meli R, Calignano A. Potential beneficial effects of butyrate in intestinal and extraintestinal diseases. World J Gastroenterol 2011;17(12):1519–28.
[38] Dizdar EA, Sari FN, Degirmencioglu H, Canpolat FE, Oguz SS, Uras N, et al. Effect of mode of delivery on macronutrient content of breast milk. J Matern Fetal Neonat Med 2014;27(11):1099–102.
[39] Perez M, Ladero V, Redruello B, Del Rio B, Fernandez L, Rodriguez JM, et al. Mastitis modifies the biogenic amines profile in human milk, with significant changes in the presence of histamine, putrescine and spermine. PLoS One 2016;11(9): e0162426.
[40] Zhang M, Caragine T, Wang H, Cohen PS, Botchkina G, Soda K, et al. Spermine inhibits proinflammatory cytokine synthesis in human mononuclear cells: a counter-regulatory mechanism that restrains the immune response. J Exp Med 1997;185 (10):1759–68.

CHAPTER SIX

Human milk oligosaccharides

"Oatmeal tastes so good on its own, you don't even need to add sugar"
Kristin Chenoweth

Key Points

1. Human milk oligosaccharides undergo very little digestion via human-derived enzymatic processes, and intestinal microbes provide the main mechanisms for their digestion.
2. Microbially digested oligosaccharides yield bioactive metabolites that play a significant role in intestinal immunoprotection as well as providing several other beneficial roles.
3. Specific microbial species such as those under the Bifidobacterium genus may play a major role in interacting with the oligosaccharides.
4. Specific human genetic variations such as "secretor status" confer the ability to provide the most protective of these oligosaccharides in milk.

Introduction

Human milk oligosaccharides (HMOs) are found in relatively high quantities in human milk. After lactose, they represent the highest concentration of carbohydrates [1,2]. Like undigestible fiber for adults, these have raised considerable interest in terms of their function and benefits. Their total concentration is estimated to be between 5 and 15 g/L in human milk samples. HMOs are not digested by the infant gastrointestinal tract, but they interact with microbes and their enzymes that can break them into smaller and highly bioactive metabolites. They are not considered as classic macronutrients written about in Chapter 2, Macronutrients

The Science of Human Milk.
DOI: https://doi.org/10.1016/B978-0-323-99913-7.00009-5

and Micronutrients, of this book. However, they are major bioactive constituents and because of recent discoveries deserve considerable attention as to their structure, function, and therapeutic potential.

The early research on human milk involved the discovery that human milk contained unique oligosaccharides [3]. Since then, there has been a growing interest in HMOs because of their putative physiologic, immunologic, and epigenetic roles, how they relate to the microbial ecosystem in human milk, and how their metabolites affect the developing infant gastrointestinal tract [2]. In addition to numerous other factors, they also engender a prebiotic role, whereby they serve as food for microbes. Indeed, not only do they serve as prebiotics, stimulating the development of a favorable microbiota, but they also act as antiadhesive (preventing microbial adhesion to the intestinal surface) and antimicrobial factors [4]. Accordingly, there is emerging evidence that HMOs are also associated with a lower risk of various types of infections [5]. They also appear to provide metabolites such as sialic acid that enhance neurocognitive development [6]. They are also thought to program the infant immune system in a manner that might protect against the development of allergies and other immune-mediated illnesses [7].

In this chapter, we discuss various aspects of HMOs including their structure, genetic variants, function, how they interact with the intestinal mucosal immune system, and their putative health benefits.

Structure

HMOs are elongation products synthetized by glycosyltransferases. These are enzymes that variably combine five different monosaccharides into more than 150 different HMO structures with different roles [8]. These five monomers are _-D-galactose (Gal), _-D-glucose (Glc), _-D-Nacetyglucosamine (GlcNAc), _-L-fucose (Fuc), and the sialic acid _-D-N-acetylneuraminic acid (Sia) [9]. Lactose is formed in the lactating mammary glands by an enzymatic system called lactose synthase complex. Elongation of lactose with these mono and disaccharides is brought about by a series of other enzymes found in the lactating mammary glands.

Biological functions of oligosaccharides are closely related to their conformation. The monomers of milk oligosaccharides from which they arise are D-glucose (Glc), D-galactose (Gal), N-acetylglucosamine (GlcNAc), L-fucose (Fuc), and sialic acid [N-acetylneuraminic acid (NeuAc)]. Most oligosaccharides carry lactose at their reducing end. An elongation is achieved by an enzymatic attachment of GlcNAc residues linked in ß1−3 or in ß1−6 linkage to a Gal residue followed by further addition of Gal in a ß-1−3 or ß-1−4 bond [10,11].

From an evolutionary perspective, HMOs are found in most mammals including the monotremes, marsupials, and eutherians (see Chapter 1: Evolution of Human Milk and Breastfeeding). Their molecular compositions vary markedly from species to species. Comparison of HMO composition from mothers who give birth to preterm versus term infants shows structural differences [10]. The stage of lactation (colostrum, transitional, and mature) is also a key parameter affecting HMO profiles.

HMOs are not only found in relatively high concentrations, but they also exhibit relatively high diversity when compared to farmed animals used for human nutrition such as cows (Table 6.1) [10]. Each of these nearly 200 oligosaccharides in human milk is distinct from one another.

Table 6.1 Olicosaccharides in human milk and cow's milk.

	Amount (g/L)	
Component	**Human milk**[a]	**Cow's milk**[b]
Lactose	55−70	40−50
Oligosaccharides		
Lacto-*N*-tetraose	0.5−1.5	Traces
Lacto-*N*-fucopentaose I	1.2−1.7	—
Lacto-*N*-fucopentaose II	0.3−1.0	—
Lacto-*N*-fucopentaose III	0.01−0.2	—
Lacto-*N*-difucohexaose I	0.1−0.2	—
NeuAc (α2−6) lactose	0.3−0.5	
		0.03−0.06
NeuAc(α2−3)lactose	0.1−0.3	
NeuAc-lacto-*N*-tetraose a	0.03−0.2	Traces
NeuAc-lacto-*N*-tetraose c	0.1−0.6	Traces
$NeuAc_2$-lacto-*N*-tetraose	0.2−0.6	Traces
Oligosaccharides (total)	5.0−8.0	Traces

[a]Data are obtained from Kunz C., Rudloff S. Biological functions of oligosaccharides in human milk. Acta Paediatr 1993;82:903−912 and Montreuil J., Mullet S. Etude des variations des constituents glucidiques du lait de femme au cours de la lactation. Bull Soc Chim Biol 1960;42:365−377.
[b]Data are obtained from Kobata A. Methods Enzymol 1972;28:262 and Parkkinen J., Finne, J. Methods Enzymol 1987;138:289.
Source: From Kunz C, Rudloff S, Baier W, Klein N, Strobel S. Oligosaccharides in human milk: structural, functional, and metabolic aspects. Annu Rev Nutr 2000;20:699−722.

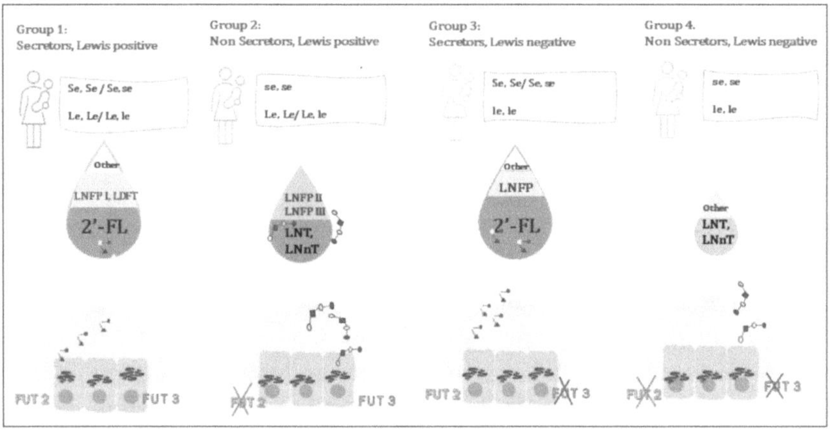

Figure 6.1 Concepts of secretor and nonsecretor status. *From Corona L, Lussu A, Bosco A, Pintus R, Cesare Marincola F, Fanos V, et al. Human milk oligosaccharides: a comprehensive review towards metabolomics. Child (Basel) 2021;8(9):804.*

This suggests the existence of many structure-specific functions, which are just beginning to be understood [11].

Maternal genetic polymorphisms are associated with structural differences in HMOs and their level of secretion (Fig. 6.1) [12]. The Lewis blood group antigen system is based upon two genes on chromosome 19: FUT3, or Lewis gene; and FUT2, or Secretor gene. Both genes are expressed in glandular epithelia including that found in the breast. The proteins produced by the FUT2 and FUT3 genes modify type I oligosaccharide chains to create Lewis antigens in blood cells. These oligosaccharide chains are like the type II chains of the ABO blood system, with a single bond in a different position. HMO profiles strongly change by maternal polymorphisms in the Lewis blood group system (FUT2 and FUT3) and duration of lactation.

Women who express at least one Secretor dominant allele are called "secretors" and synthetize high amounts of 20-FL. Women who express two recessive alleles are called "nonsecretors." Since they do not possess an active enzyme, their milk presents a significantly lower amount of 20-FL [13]. This also explains why total HMO concentration in "nonsecretor's" milk is 35%–40% lower than in "secretor's."

The prevalence of FUT 2 negative genotypes varies across different geographic regions. Up to 30% of African Americans and Latin Americans show negative genotypes, whereas the negative genotype in European populations is extremely rare [14].

Environmental and maternal factors may also contribute to HMO variability, but these are not currently well understood. It is thought that secretor and Lewis status defined by specific *FUT2* and *FUT3* SNPs may be proxies for specific individual HMOs and milk group variabilities [12].

Function

Some HMOs have redundant roles, and some may have synergistic functional effects. These are not well understood and require more than simply research related to associations. It is known, however, that HMOs drive the development of the microbiome, strengthen gut barrier function, support immune competence development, and modulate the developing nervous system and cognitive development. Several of these factors are accomplished via the production of the interaction of the HMOs and microbes and the production of metabolites resulting from the metabolic interactions [1]. We will next elaborate on these functions in more detail.

HMOs and the intestinal microbiome

As described in the previous chapter various bioactive components of human milk are numerous. It is thus unlikely that any one bioactive component will play a major role by itself in the overall physiology of the developing gastrointestinal tract in infants. One area of interest considers the synergy of some of the activities of several of the gastrointestinal tract bioactive components, oligosaccharides, and the microbial milieu of the intestine. Breastfed infants show a high number of relative abundances of bifidobacteria family members during the early breastfeeding [15]. The role of this high abundance of bifidobacteria during early life as it relates to the HMOS appears to be important for intestinal development and is just beginning to be understood. The presence of bifidobacteria is highly related to that of the HMOs present in milk. It is thought that bifidobacteria species are highly equipped to take advantage of the HMOs found in breast milk. Their unique glycan foraging capability allows Bifidobacteria to utilize HMOs by (1) secretion of glycosidases that degrade HMOs which then are followed by the uptake of the resultant monomers by the Bifidobacteria, and (2) their expression of HMO transporters that can further allow for degradation of these HMOs internally [16].

Infant gut utilization of HMOs by microbes has been related to lower inflammatory markers and this has also been associated with Bifidobacterium subspecies-derived metabolites such as indole acetate. It is highly suspected that these activities are very relevant to the development of immune competence.

HMOs and growth

HMOs have also been investigated in relation to infant growth [17]. Studies linking HMOs and growth suggest that specific oligosaccharides may act via the microbiota in affecting infant growth. Most of these have been observational studies that exhibit associations and are not consistent. One hypothesis that suggests that HMO intake may increase food efficiency through microbiota-related processes such as breakdown to short-chain fatty acids that can be utilized as energy substrates as well as having numerous other bioactive functions has been raised. Another hypothesis is that HMOs affect food responsiveness and appetite through a microbiome-driven enteroendocrine central nervous system process. However, at this juncture these hypotheses have not been adequately tested using controlled and randomized conditions or other techniques such as multiomic integrations that could result in a better delineation of mechanistic pathways.

HMOs and immunity

Development of the immune system is dependent on the microbial composition of the gastrointestinal tract [18]. The role of HMOs for term infants has yet to be established. However, there have been several studies suggesting an association in preterm infants and animal models between HMO feedings and intestinal injury. Although it has been suggested that these studies may be highly relevant to neonatal necrotizing enterocolitis

(NEC), they remain suspect since these animal models poorly represent the problem seen in preterm human infants. Studies of oligosaccharide administration in preterm infants have been underpowered and thus raise questions about whether these studies support the contention that these agents may protect against intestinal injuries such as those referred to as "NEC" seen in preterm infants.

HMOs and neurodevelopment

Many functional processes in the brain depend on sialic acid bound to proteins and glycolipids. There is less high sialic acid demand during early development. Breast milk contains high levels of sciatic acid primarily in the form of HMOs. However, studies in animals support that most dietary sialic acid is catabolized to precursors that are not used directly as sialic acid. Furthermore, there are other metabolic pathways wherein there is uptake and incorporation of sialic acid via a salvage pathway and thus may still play a role in brain and cognitive development. Considerable work needs to be done to determine the relationship between HMOs and neurodevelopment [19].

Conclusion

HMOs, from a chemical structure perspective, represent numerous features that are present on mucosal and cell surface glycans that play an important modulatory role in various physiological interactions. From a functional perspective, there is evidence that suggests that they may play important roles in neurodevelopment, immunological activity, and growth. It appears that the oligosaccharides require synergy with certain bacteria such as the Bifidobacteria to be functionally relevant. At this juncture, the evidence supporting the addition of HMOs to formulas is insufficient to indicate a beneficial role for this practice. The matrix of human milk is very complex and the function changes dynamically over time, which is very difficult to mimic with formulas with additives such as oligosaccharides.

References

[1] Corona L, Lussu A, Bosco A, Pintus R, Cesare Marincola F, Fanos V, et al. Human milk oligosaccharides: a comprehensive review towards metabolomics. Child (Basel) 2021;8(9):804.
[2] Bode L. Human milk oligosaccharides: every baby needs a sugar mama. Glycobiology 2012;22(9):1147–62.
[3] Polonovski M, Montreuil J. Chromatographic study of the polyosides of human milk. C R Hebd Seances Acad Sci 1954;238(23):2263–4.
[4] Craft KM, Townsend SD. Mother knows best: deciphering the antibacterial properties of human milk oligosaccharides. Acc Chem Res 2019;52(3):760–8.
[5] Duijts L, Jaddoe VW, Hofman A, Moll HA. Prolonged and exclusive breastfeeding reduces the risk of infectious diseases in infancy. Pediatrics 2010;126(1):e18–25.
[6] Mudd AT, Fleming SA, Labhart B, Chichlowski M, Berg BM, Donovan SM, et al. Dietary sialyllactose influences sialic acid concentrations in the prefrontal cortex and magnetic resonance imaging measures in corpus callosum of young pigs. Nutrients. 2017;9(12):1297.
[7] Doherty AM, Lodge CJ, Dharmage SC, Dai X, Bode L, Lowe AJ. Human milk oligosaccharides and associations with immune-mediated disease and infection in childhood: a systematic review. Front Pediatr 2018;6:91.
[8] Urashima T, Hirabayashi J, Sato S, Kobata A. Human milk oligosaccharides as essential tools for basic and application studies on galectins. Trends Glycosci Glycotechnol 2018;30(172):SE51–65.
[9] Palmeira P, Carneiro-Sampaio M. Immunology of breast milk. Rev Assoc Med Bras (1992) 2016;62(6):584–93.
[10] Kunz C, Rudloff S, Baier W, Klein N, Strobel S. Oligosaccharides in human milk: structural, functional, and metabolic aspects. Annu Rev Nutr 2000;20:699–722.
[11] Smilowitz JT, Lebrilla CB, Mills DA, German JB, Freeman SL. Breast milk oligosaccharides: structure-function relationships in the neonate. Annu Rev Nutr 2014;34:143–69.
[12] Lefebvre G, Shevlyakova M, Charpagne A, Marquis J, Vogel M, Kirsten T, et al. Time of lactation and maternal fucosyltransferase genetic polymorphisms determine the variability in human milk oligosaccharides. Front Nutr 2020;7:574459.
[13] Blank D, Dotz V, Geyer R, Kunz C. Human milk oligosaccharides and Lewis blood group: individual high-throughput sample profiling to enhance conclusions from functional studies. Adv Nutr (Bethesda, MD) 2012;3(3):440S–9S.
[14] Wang A, Koleva P, du Toit E, Geddes DT, Munblit D, Prescott SL, et al. The milk metabolome of non-secretor and Lewis negative mothers. Front Nutr 2020;7:576966.
[15] Walker WA. Initial intestinal colonization in the human infant and immune homeostasis. Ann Nutrit Metab 2013;63(Suppl 2):8–15.
[16] Garrido D, Kim JH, German JB, Raybould HE, Mills DA. Oligosaccharide binding proteins from *Bifidobacterium longum* subsp. infantis reveal a preference for host glycans. PLoS One 2011;6(3):e17315.
[17] Eriksen KG, Christensen SH, Lind MV, Michaelsen KF. Human milk composition and infant growth. Curr Opin Clin Nutr Metab Care 2018;21(3):200–6.
[18] Plaza-Díaz J, Fontana L, Gil A. Human milk oligosaccharides and immune system development. Nutrients. 2018;10(8):1038.
[19] Oliveros E, Martín M, Torres-Espínola F, Segura-Moreno M, Ramírez M, Santos A, et al. Human milk levels of 2-fucosyllactose and 6-sialyllactose are positively associated with infant neurodevelopment and are not impacted by maternal BMI or diabetic status. J Nutr Food Sci 2021;4(1):100024.

CHAPTER SEVEN

Epigenetics of human breast milk

First Law: In every animal which has not passed the limit of its development, a more frequent and continuous use of any organ gradually strengthens, develops and enlarges that organ, and gives it a power proportional to the length of time it has been so used; while the permanent disuse of any organ imperceptibly weakens and deteriorates it, and progressively diminishes its functional capacity, until it finally disappears.

Second Law: All the acquisitions or losses wrought by nature on individuals, through the influence of the environment in which their race has long been placed, and hence through the influence of the predominant use or permanent disuse of any organ; all these are preserved by reproduction to the new individuals which arise, provided that the acquired modifications are common to both sexes, or at least to the individuals which produce the young.

Jean-Baptiste Lamarck, *Zoological Philosophy*

Key Points

1. Components of human milk along with microbial interactions can regulate infants' gene expression via epigenetic mechanisms. These can lead to profound changes in phenotype.
2. Some of the major epigenetic mechanisms related to human milk are DNA methylation, histone modification, and short interfering RNA (siRNA) gene silencing.
3. Human milk and its interaction with the infant and the environment can alter these mechanisms based on individual exposures.

Introduction

The sequence of nucleotides in an organism's DNA encompasses the basic code for subsequent phenotypic development. However,

The Science of Human Milk.
DOI: https://doi.org/10.1016/B978-0-323-99913-7.00012-5

environmental influences play a major role as to how and when genetic information is expressed. Epigenetics refers to changes into how the genome is expressed that have functional relevance but do not involve changes in the gene sequence.

Human milk, especially when derived directly from the infant's own mother provides highly personalized nutrition for the infant. Various nutrients can alter the infant's gene expression, hence amplifying or turning off various gene functions via epigenetic mechanisms. Nutritional epigenetics relates to how genes may be expressed depending on the effects of nutrients in an individual. Nutrition in early life during critical periods of development is known to modulate gene expression. How these genes are modulated during early development is known to have effects on subsequent health. Human breast milk and its components are known to prevent several acute and chronic diseases. Many of these effects relate to epigenetic mechanisms.

Current nutrition recommendations are targeted toward estimated average nutrient requirements for a population. They aim to meet the needs of most individuals within a population [1]. However, these are not precision based and will not meet specific nutritional needs of many individuals. In the case of specific genetic polymorphisms, personalized recommendations may be needed [2]. Nutrients are able to regulate gene expression via epigenetic mechanisms [3,4]. These processes determine when, where, and how the information contained in specific genes is expressed. Thus profound changes in phenotype may result from epigenetic regulation of genes. Major epigenetic regulatory processes include DNA methylation, histone modification, chromatin remodeling, and microRNAs (Fig. 7.1).

Many studies that relate early-life nutrition to epigenetic regulation of genes have focused on DNA methylation (see Fig. 7.1) [5,6]. Methylation tags the cytosine nucleotide within the genome via a DNA methyltransferase that subsequently forms 5-methylcytosine. Most DNA methylation occurs within CpG dinucleotides located in the promoter region of a gene [6]. Increased methylation that binds methyl-CpG binding proteins blocks transcription and thus is usually associated with decreased transcription. Therefore epigenetics, by influencing the expression of genes, controlled by the intake of different types of food and other environmental factors during early life, can modify the risk for the development of many common diseases [5,7].

Maternal malnutrition during pregnancy results in a retardation of fetal growth. It also causes modifications of the expression of biochemical

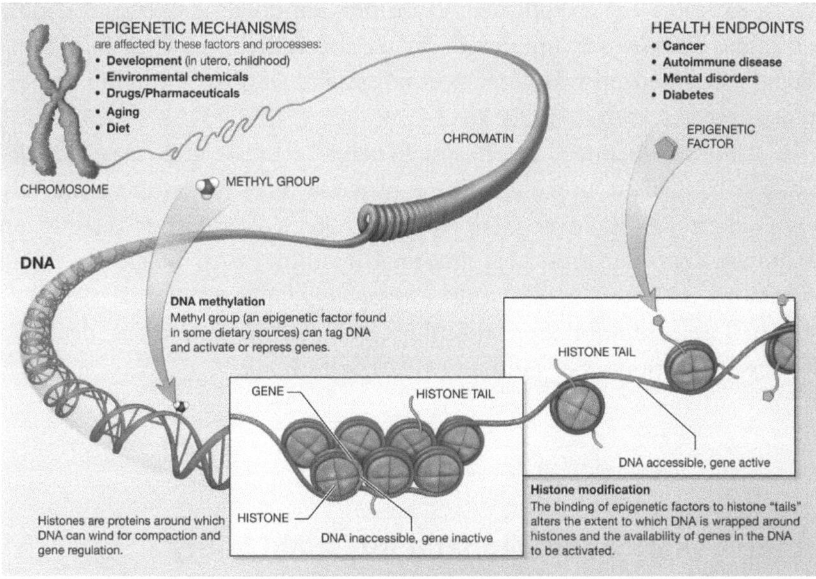

Figure 7.1 Epigenetic mechanisms and health endpoints. *From https://en.wikipedia.org/wiki/Epigenetics.*

mechanisms related to the endocrinological and metabolic control [8]. As an example, offspring of mothers who have undergone a protein-restricted diet from conception throughout pregnancy exhibit an altered metabolic phenotype. They exhibit several features of human cardio-metabolic disease, including hypertension, obesity, dyslipidemia, vascular dysfunction, and impaired glucose homeostasis [9]. This is sometimes referred to as the "metabolic syndrome" or "Barker hypothesis" [10].

Several studies, focused on individuals exposed to famine in utero which occurred in the Netherlands during the winter of 1944, provide evidence for epigenetic mechanisms causing metabolic syndrome. Individuals whose mothers experienced famine periconceptually and in the first trimester of pregnancy were born with low birth weight compared with unexposed individuals. When they became adults, they exhibited an increased risk of obesity and cardiovascular disease [11].

Many of the infants born to mothers who experienced famine were subsequently exposed to adequate food intake and exhibited catch-up growth. Infants born preterm, who have a reduced fat mass, especially those who are formulafed, who exhibit early catch-up growth show an increased risk of cardio-metabolic disease in later life, including obesity [12]. The exact mechanisms underlying how early nutrition with subsequent catch-up growth may cause programming of risk of noncommunicable diseases are unknown. It has been hypothesized that these changes are associated with altered development of organ structure or persistent alteration at the cellular level [13].

In a likewise manner, the health benefits of breast milk may partially be due to epigenetic processes. These processes may be complex and rely on the interrelationship of environmental effects of diet not directly on the human cells, but rather via intestinal microbes, their metabolites, and interactions with various host cell immunologic and proteomic responses [14,15]. These multiomic relationships are discussed in more detail in Chapter 5, Multiomics of Human Milk, of this book.

Epigenetics and neonatal intestinal injury

Breast versus bottle feeding with commercial formula may strongly influence the short-term composition of infant's gut microbiota [16]. At 6 months, infants have developed an individualized microbiota signature that lasts throughout their lifetime. Appropriate colonization is influenced by various nutritional factors including the effect of breast milk oligosaccharides. These undigestible complex carbohydrates (e.g., 8% of total calories) function as prebiotics, providing substrate for certain types of microbes to produce short-chain fatty acids and other potentially beneficial metabolites. This enhances the proliferation of health-promoting bacteria, such as *Bifidobacteria* and *Lactobacillus*. There is a known association between levels of protective secretory IgA in intestinal secretions and the number of *Bifidobacteria* in the gut at 1 month of age [17].

Human breast milk suppresses the interleukin (IL) 1-β-induced activation of the IL-8 gene promoter in human intestinal cells by inhibiting the activation of nuclear factor kappa B (NF-κB) [18]. NF-κB is a very important protein transcription factor that is a regular of innate immunity. It serves as a link between pathogenic signals and cellular immune responses that provide resistance to invading pathogens. Studies in cell culture have demonstrated that many species of commensal bacteria may reduce the inflammatory response by inhibiting the nuclear factor kappa-light-chain enhancer of activated B cells (NF-κB) [19]. Such a reduction in the inflammatory response is very important since it acts as a counterbalance to the proinflammatory responses that may be extremely brisk in infants. This balance of pro- and antiinflammatory signaling is critical in maintaining normal intestinal functions and maintains intestinal homeostasis in breastfed infants.

Research on the immunomodulatory role of human breast milk through regulation of gene expression has implicated that lactoferrin protective properties may play a role. As described in Chapter 3, Bioactive Components of Human Milk, lactoferrin is an abundant breast milk protein that binds proinflammatory bacterial DNA CpG in extracellular compartments [20]. This binding is thought to inhibit the CpG-motif DNA-induced activation of NF-κB-regulated genes, such as IL-8 and IL-12 in B cells [20]. Bacterial DNA CpG motifs may be present in the lamina propria and Peyer's patches due to lysis of enteropathogens, and human milk lactoferrin may modulate immune responses in the lymphoid follicles of the infant's intestine through its action on these enteropathogens [20].

Effects of breastfeeding on subsequent health

Breastfeeding is associated with a reduction in the risk of childhood inflammatory bowel disease and celiac disease of 31% and 52%, respectively. This effect was most marked in infants who were exclusively breastfed at the time of gluten exposure [21]. This protective role is postulated to result from the interaction of underlying genetic susceptibility, the immunomodulating effect of human milk, and the different patterns of gut microbiota composition in breastfed versus formulafed infants [22,23].

As previously mentioned, breastfeeding, compared with formula feeding, has been associated with a lower risk of metabolic syndrome. Lower protein and energy content in breast milk than formula, jointly with higher content of long-chain polyunsaturated fatty acids (LCPUFA), cholesterol, and nondigestible carbohydrate, may act as a substrate for beneficial strains of bacteria to grow in the intestine. Examples of such bacteria include Bifidobacteria and Lactobacillus species. This, in turn, may act synergistically with better hunger and satiety self-regulation of breastfed infants (possibly modulated by some hormones and hormone-like compounds, e.g., ghrelin and leptin) and thus lead to healthier outcomes [24].

Cholesterol regulation

Breastfeeding is associated with increased mean total cholesterol and LDL cholesterol levels in infancy but lower levels in adulthood [25]. The high cholesterol content of breast milk acting via feedback mechanisms may well be responsible. High cholesterol intake in infancy may reduce endogenous cholesterol synthesis, probably by downregulation of hepatic hydroxymethyl glutaryl coenzyme A (HMGCoA) reductase [25].

Short-chain fatty acids

In Chapter 6, Human Milk Oligosaccharides, the concept about short-chain fatty acids produced by interaction of bacteria and human milk oligosaccharides was introduced. There are many functions of these short-chain fatty acids, especially the four-carbon chain fatty acid, butyrate. Butyrate was able to exert several immunomodulatory functions. In a study designed to determine whether butyrate concentrations in human milk are able to modulate mechanisms of immune tolerance, butyrate concentrations were assessed in 109 healthy women in order to evaluate its effect on tolerogenic mechanisms in both in vivo and in vitro models. It was found that the butyrate concentration upregulated the expression of several biomarkers of gut barrier integrity and tolerogenic cytokines. Pretreatment with butyrate also significantly reduced allergic responses in three animal models of allergy. It was thought that this response was partially through epigenetic mechanisms add suggested the importance of butyrate as a pivotal human milk compound that can protect against food allergy [26].

In another study that evaluated the role of short-chain fatty acids in the developmental programming of the immune system the epigenetic effects involved quantification of human milk short-chain fatty acid levels in human milk samples from atopic versus nonatopic mothers [27]. Samples from atopic mothers had significantly lower concentrations of acetate and butyrate than those of nonatopic mothers. The authors hypothesized that reduced exposure to human milk short-chain fatty acids in early life may program for atopic disease later.

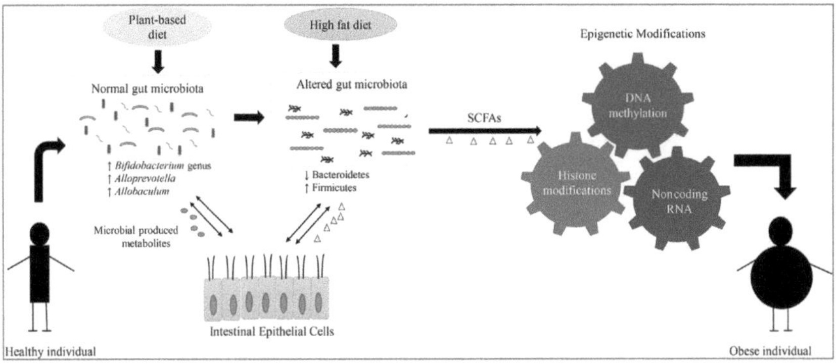

Figure 7.2 Interaction between fat intake, gut microbiome, epigenetics, and obesity. *From Sharma M, Li Y, Stoll ML, Tollefsbol TO. The epigenetic connection between the gut microbiome in obesity and diabetes. Front Genet 2020;10, 1329.*

Similar epigenetic mechanisms have been invoked to explain a predisposition to obesity with different dietary intakes such as plant or high-fat diets. The scheme in Fig. 7.2 below shows an interplay between the diet, gut microbiome, epigenetic mechanisms, and the development of obesity.

The types of microorganisms arising from a plant-based diet are different than those resulting from a high-fat animal diet. The metabolites produced by these microorganisms differ and include the short-chain fatty acids. These interact with the epithelial cells of the host which induce specific epigenetic alterations such as histone modification, noncoding RNA modifications, and DNA methylation which in turn dispose to the development of obesity [28]. It is conceivable that dietary habits in the lactating mother may alter her intestinal microbiome, hence influencing the metabolites that are absorbed into her bloodstream and hence into her milk. These concepts remain understudied areas and require confirmation with additional research.

Other components of human milk with epigenetic potential

Possible epigenetic effect of human breast milk components on a child's health outcomes is summarized in Table 7.1

Table 7.1 Epigenetic effect of human breast milk components on the child's health outcomes.

Human milk component	**Prevent**	**Gene (expression)**
Lactoferrin	NEC disorders of immune system	NF-kB (reduced)[a]
Prostaglandin J	Obesity and related-disorders	PPARγ (increased)[b]
LCPUFA $n-3$	NAFLD	Liver lipogenic and cholesterol biosynthesis enzymes (reduced)[a]
	Progression of NAFLD	PPAR α and γ (increased)[b]
	High blood total cholesterol in adulthood	HMGCoA reductase (reduced)[a]
Cholesterol content	High blood total cholesterol in adulthood	HMGCoA reductase (reduced)[b]
Undigestible oligosacch arides	Gut dysbiosis and related alterations (NEC, infectious diseases, disorders of immune system, obesity, and linked disorders)	Action on expression of different genes (e.g., NF-κB)[b]

[a]Proved in vitro and/or animals.
[b]Hypothesized in humans.
Source: From Verduci E, Banderali G, Barberi S, Radaelli G, Lops A, Betti F, et al. Epigenetic effects of human breast milk. Nutrients 2014;6(4):1711–24.

Summary

The different epigenetic mechanisms that correlate with the beneficial effects of breast milk feeding remain unclear. Breast milk, modulating gene expression without changing the nucleotide sequence of DNA, might positively modify the phenotype and the outcome even if there is a genetic predisposition for the development of diseases. These alterations may result from regulation through diet, environment, and various stressors. This can be done via alterations of the microbiome in the maternal or infant gastrointestinal tract, which in turn affects the metabolic milieu of the infant such as an increase in methyl components which can silence DNA transcription or alter the histone configuration. Short-chain fatty acids produced by these microbes with exposure to certain substrates can also have similar epigenetic effects. Unlike genetic mutations, it is important to remember that if discovered early enough these epigenetic alterations may be reversible, offering the potential for both prevention of subsequent disease or therapy.

References

[1] EFSA Panel on Dietetic Products Nutrition, and Allergies. Scientific opinion on establishing food-based dietary guidelines. EFSA J 2010;8(3):1460.
[2] Hurlimann T, Menuz V, Graham J, Robitaille J, Vohl MC, Godard B. Risks of nutrigenomics and nutrigenetics? What the scientists say. Genes Nutr 2014;9(1):370.
[3] Ho E, Zempleni J. Overview to symposium "Nutrients and epigenetic regulation of gene expression". J Nutr 2009;139(12):2387–8.
[4] Stover PJ, Caudill MA. Genetic and epigenetic contributions to human nutrition and health: managing genome-diet interactions. J Am Diet Assoc 2008;108(9):1480–7.
[5] Liotto N, Miozzo M, Giannì ML, Taroni F, Morlacchi L, Piemontese P, et al. Early nutrition: the role of genetics and epigenetics. Pediatr Med Chir 2009;31(2):65–71.
[6] Tammen SA, Friso S, Choi SW. Epigenetics: the link between nature and nurture. Mol Asp Med 2013;34(4):753–64.
[7] Cutfield WS, Hofman PL, Mitchell M, Morison IM. Could epigenetics play a role in the developmental origins of health and disease? Pediatr Res 2007;61(5 Pt 2):68R–75R.
[8] Waterland RA, Michels KB. Epigenetic epidemiology of the developmental origins hypothesis. Annu Rev Nutr 2007;27:363–88.
[9] Verduci E, Banderali G, Barberi S, Radaelli G, Lops A, Betti F, et al. Epigenetic effects of human breast milk. Nutrients 2014;6(4):1711–24.
[10] Barker D, Eriksson J, Forsen T, Osmond C. Fetal origins of adult disease: strength of effects and biological basis. Int J Epidemiol 2002;31(6):1235–9.
[11] Heijmans BT, Tobi EW, Stein AD, Putter H, Blauw GJ, Susser ES, et al. Persistent epigenetic differences associated with prenatal exposure to famine in humans. Proc Natl Acad Sci U S A 2008;105(44):17046–9.
[12] Barker DJP. Sir Richard Doll Lecture. Developmental origins of chronic disease. Public Health 2012;126(3):185–9.
[13] Koletzko B, Brands B, Poston L, Godfrey K, Demmelmair H. Early nutrition programming of long-term health. Proc Nutr Soc 2012;71(3):371–8.
[14] Shenhav L, Azad MB. Using community ecology theory and computational microbiome methods to study human milk as a biological system. mSystems. 2022;7(1): e0113221.
[15] Du C, Zhao Y, Wang K, Nan X, Chen R, Xiong B. Effects of milk-derived extracellular vesicles on the colonic transcriptome and proteome in murine model. Nutrients 2022;14(15):3057.
[16] Fanaro S, Chierici R, Guerrini P, Vigi V. Intestinal microflora in early infancy: composition and development. Acta Paediatr (Oslo, Norway: 1992) Suppl 2003;91 (441):48–55.
[17] Sjögren YM, Tomicic S, Lundberg A, Böttcher MF, Björkstén B, Sverremark-Ekström E, et al. Influence of early gut microbiota on the maturation of childhood mucosal and systemic immune responses. Clin Exp Allergy 2009;39(12):1842–51.
[18] Minekawa R, Takeda T, Sakata M, Hayashi M, Isobe A, Yamamoto T, et al. Human breast milk suppresses the transcriptional regulation of IL-1beta-induced NF-kappaB signaling in human intestinal cells. Am J Physiol Cell Physiol 2004;287(5): C1404–11.
[19] Chen AC, Chung MY, Chang JH, Lin HC. Pathogenesis implication for necrotizing enterocolitis prevention in preterm very-low-birth-weight infants. J Pediatr Gastroenterol Nutr 2014;58(1):7–11.
[20] Mulligan P, White NR, Monteleone G, Wang P, Wilson JW, Ohtsuka Y, et al. Breast milk lactoferrin regulates gene expression by binding bacterial DNA CpG motifs but not genomic DNA promoters in model intestinal cells. Pediatr Res 2006;59(5):656–61.

[21] Eidelman A.I., Schanler R.J., Johnston M., Landers S., Noble L., Szucs K., et al. Breastfeeding and the use of human milk. Pediatrics 2012;129 (3): e827–e841.
[22] Moloney RD, Desbonnet L, Clarke G, Dinan TG, Cryan JF. The microbiome: stress, health and disease. Mamm Genome 2014;25(1–2):49–74.
[23] Weng M, Walker WA. The role of gut microbiota in programming the immune phenotype. J Dev Orig Health Dis 2013;4(3):203–14.
[24] Agostoni C, Baselli L, Mazzoni MB. Early nutrition patterns and diseases of adulthood: a plausible link? Eur J Intern Med 2013;24(1):5–10.
[25] Owen CG, Whincup PH, Kaye SJ, Martin RM, Davey Smith G, Cook DG, et al. Does initial breastfeeding lead to lower blood cholesterol in adult life? A quantitative review of the evidence. Am J Clin Nutr 2008;88(2):305–14.
[26] Paparo L, Nocerino R, Ciaglia E, Di Scala C, De Caro C, Russo R, et al. Butyrate as a bioactive human milk protective component against food allergy. Allergy 2021;76(5):1398–415.
[27] Stinson LF, Gay MCL, Koleva PT, Eggesbø M, Johnson CC, Wegienka G, et al. Human milk from atopic mothers has lower levels of short chain fatty acids. Front Immunol 2020;11:1427.
[28] Sharma M, Li Y, Stoll ML, Tollefsbol TO. The epigenetic connection between the gut microbiome in obesity and diabetes. Front Genet 2020;10:1329.

CHAPTER EIGHT

Human milk exosomes/ extracellular vesicles

Plough ahead as a steamer does, rough or smooth - rain or shine. To carry your cargo and make your port is the point.

Maltbie Davenport Babcock

Key Points

1. Exosomes are extracellular vesicles generated to carry nucleic acids, proteins, lipids, and metabolites. They are mediators of near and long-distance intercellular communication in health and disease.
2. With transfer of breast milk from mother to infant, exosomes/EVs participate in intercellular communication by fusing with the infant recipient cell membranes and delivering the parent cell molecular cargo to the infant target cell.
3. Human breast milk exosomes are resilient to digestion and can be endocytosed by infant intestinal epithelial cells, indicating that these have the therapeutic potential if mimicked in commercial formulas.

Introduction

Most cells within the body need to communicate to perform their function. One method is using exosomes/extracellular vesicles (EVs). These are small particles which are produced and extruded by cells throughout the body. They are loaded with active regulatory and stimulatory molecules from the parent cell including microRNAs (miRNAs) and enzymes. Recent research has suggested many bioactive compounds in

The Science of Human Milk.
DOI: https://doi.org/10.1016/B978-0-323-99913-7.00005-8

milk, such as nucleic acids, cannot be directly transferred from the mother to the child. The potential for degradation of these compounds is high. Digestion of these compounds via the infant intestinal tract is high. As protective intermediate carriers of these compounds, exosomes/EVs provide a protective mechanism so that these compounds can be carried as cargo to the recipient cells without being digested. Recipient cells utilize the cargo contained in these particles for various biologic responses. Breast milk, known for years to have beneficial health effects, contains a population of EVs/exosomes, which mediate its therapeutic effects.

Their various functions are summarized in Fig. 8.1.

Exosomes are extracellular vesicles. They can be generated by all cells and transport "cargo" such as nucleic acids, proteins, lipids, and metabolites from one body site to another. In milk, they act to transport such cargo from the mother to the infant. They thereby mediate intercellular communication in health and disease.

These small membrane-enclosed particles may originate from several tissues including mammary tissue and are found in most human biofluids including milk. They are heterogeneous and exhibit a range in size. Microvesicles range in size between 50 and 1000 nm and are released via

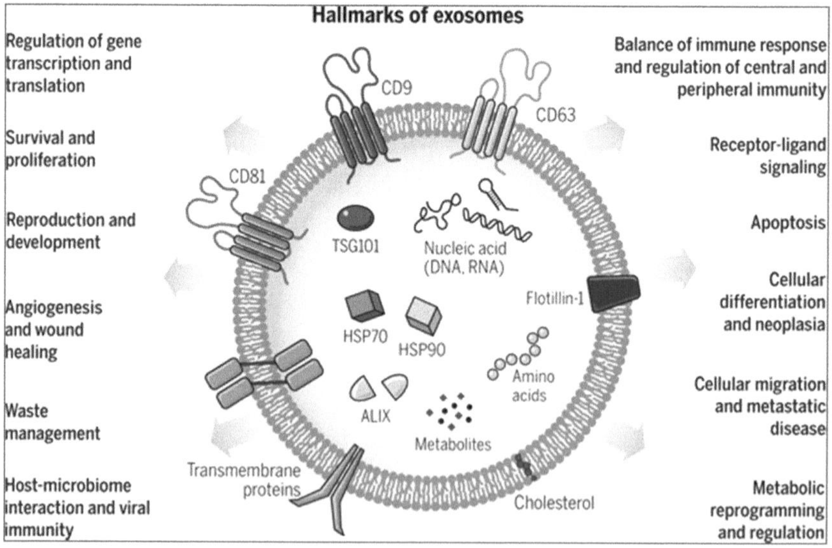

Figure 8.1 Exosomes: carriers of vulnerable cargo between mother and infant via breast milk. *From Kalluri R, LeBleu VS. The biology, function, and biomedical applications of exosomes. Science 2020;367(6478):eaau6977. https://doi.org/10.1126/science.aau6977. [1].*

blebbing of the plasma membrane. Exosomes are 30–150 nm in size. They are produced inside the cell from the endosomal network which consists of intracellular organelles that regulate trafficking of proteins and lipids [2].

There are also apoptotic bodies that are typically greater than 1000 nm in size and are created by blending of the plasma membranes from apoptotic. Milk EVs are largely made-up of the microvesicles and the exosomes, are released by the luminal epithelium of the breast cells, and are taken up by the recipient intestinal cells of the infant. These EVs can induce changes in gene expression and may play a role in immunity of the newborn [3,4].

Exosomes are initially intraluminal vesicles, located in intracellular multivesicular bodies. These multivesicular bodies usually arise from the Golgi apparatus and contain multiple internal vesicles, the endosomes. As these bodies, with their own cell membrane, directly fuse with the parent cellular membrane, the smaller vesicles (exosomes) are released into the luminal space [5]. Microvesicles, which can be considerably larger than exosomes, directly fuse with the cellular membrane and bud from the cell [6].

Both exosomes and microvesicles carry an assortment of molecules from the parent cell, including proteins, lipids, messenger RNA, and miRNAs [7,8].

With transfer of breast milk from mother to infant, these exosomes and microvesicles participate in intercellular communication by fusing with the infant recipient cell membranes and delivering the parent cell molecular cargo to the recipient target cell (Fig. 8.2) [9].

Both exosomes and microvesicles influence innate and adaptive immunity. They can modulate the NF-kB inflammatory signal transduction pathway, T-cell priming and activation, as well as development of the nervous system by promoting neuronal differentiation (Fig. 8.2A and B) [1,10,11].

Therapeutic potential

The identification of exosomes and microvesicles, referred to collectively as EVs, as intercellular communicators and the potential of their utilization from human milk have been suggested as having far-reaching effects on human health. The diverse molecules that compose EV cargo are the primary effectors of these therapeutic functions, and they can be surface-bound or found within the vesicle itself.

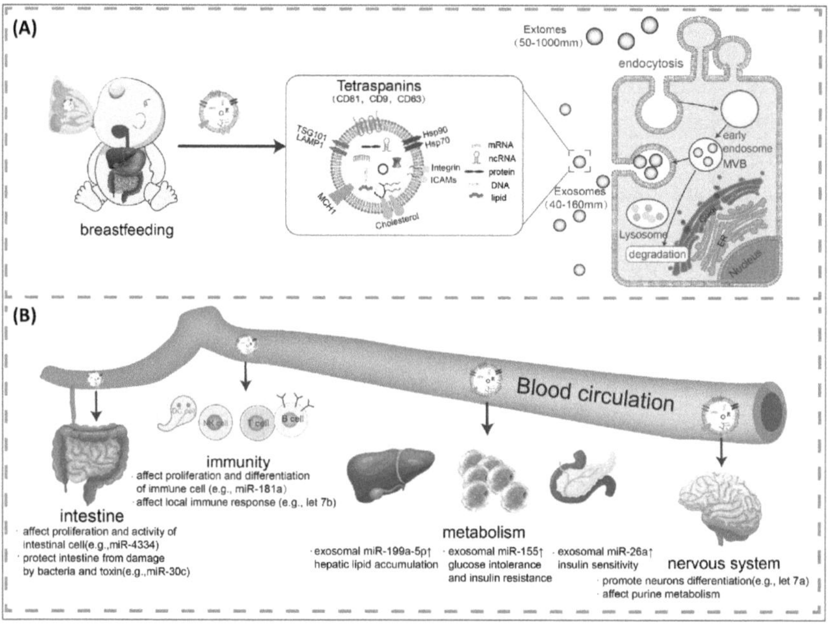

Figure 8.2 Synthesis, secretion, transport, and effects of milk-derived EVs/exosomes. *EV*, Extracellular vesicles.

The internal cargo of EVs, including enzymes and miRNAs can impart therapeutic function on recipient cells upon delivery. The miRNAs in human milk are 18–22 nucleotide noncoding RNAs. They can regulate gene expression posttranscriptionally [12].

There have been major advances in exosome research over the past decades [1]. One such innovation is the ability to engineer exosomes/EVs for delivery of specific miRNA cargo [13]. Imported through exosomes, miRNAs are highly stable. They are able to withstand the chemical and physical stresses of the intestinal tract and the bloodstream, supporting their case as potentially therapeutic tools [14,15].

Previous chapters in this book have discussed various components of human breast milk that are highly bioactive and relate to the health benefits. Human breast milk exosomes are resistant to digestion in the upper gastrointestinal tract and can be endocytosed by intestinal epithelial cells. Thus breast milk exosomes exhibit qualities of a potential therapeutic.

Research is underway to characterize the breast milk EV miRNA profile. The composition of highly abundant EV-miRNAs is similar between term and preterm milk. Both can be internalized by intestinal epithelial cells [16]. These miRNAs can then affect gene transcription within the cells.

The question about age of the milk as it affects miRNA composition has been raised. In a panda model, colostrum contained miRNAs more closely associated with immune regulation and response [17]. Interestingly, human milk collected within the first week of birth contains more exosomes compared to milk collected by the baby's second month of life [18]. The clinical implications of this finding remain unclear.

Summary

We have summarized recent research in the structure and function of exosomes/EVs, derived from BM. These are natural particles that play an important role in cell-to-cell communication by transferring RNAs, lipids, and proteins from donor cells to recipient cells. Exosomes and their cargo resist degradation in the gastrointestinal tract and during processing of milk. They can be absorbed and thus accumulate in tissues following oral administration. They also cross the blood–brain barrier. There is considerable interest in exploring milk exosomes as novel bioactive food compounds and for delivering drugs to diseased tissues. Breast milk contains large quantities of exosomes and it is thus of major interest to evaluate how these particles exert their benefits to the infants.

References

[1] Kalluri R, LeBleu VS. The biology, function, and biomedical applications of exosomes. Science 2020;367(6478):eaau6977. Available from: https://doi.org/10.1126/science.aau6977.

[2] de la Torre Gomez C, Goreham RV, Bech Serra JJ, Nann T, Kussmann M. "Exosomics"—a review of biophysics, biology and biochemistry of exosomes with a focus on human breast milk. Front Genet 2018;9:92. Available from: https://doi.org/10.3389/fgene.2018.00092.

[3] Baier SR, Nguyen C, Xie F, Wood JR, Zempleni J. MicroRNAs are absorbed in biologically meaningful amounts from nutritionally relevant doses of cow milk and affect gene expression in peripheral blood mononuclear cells, HEK-293 kidney cell cultures, and mouse livers. J Nutr 2014;144(10):1495–500. Available from: https://doi.org/10.3945/jn.114.196436.

[4] Zhou Q, Li M, Wang X, et al. Immune-related microRNAs are abundant in breast milk exosomes. Int J Biol Sci 2012;8(1):118–23. Available from: https://doi.org/10.7150/ijbs.8.118.

[5] Pan BT, Teng K, Wu C, Adam M, Johnstone RM. Electron microscopic evidence for externalization of the transferrin receptor in vesicular form in sheep reticulocytes. J Cell Biol 1985;101(3):942–8. Available from: https://doi.org/10.1083/jcb.101.3.942.
[6] Heijnen HF, Schiel AE, Fijnheer R, Geuze HJ, Sixma JJ. Activated platelets release two types of membrane vesicles: microvesicles by surface shedding and exosomes derived from exocytosis of multivesicular bodies and alpha-granules. Blood 1999; 94(11):3791–9.
[7] Haraszti RA, Didiot MC, Sapp E, et al. High-resolution proteomic and lipidomic analysis of exosomes and microvesicles from different cell sources. J Extracell Vesicles 2016;5:32570. Available from: https://doi.org/10.3402/jev.v5.32570.
[8] Waldenström A, Gennebäck N, Hellman U, Ronquist G. Cardiomyocyte microvesicles contain DNA/RNA and convey biological messages to target cells. PLoS One 2012;7(4):e34653. Available from: https://doi.org/10.1371/journal.pone.0034653.
[9] Montecalvo A, Larregina AT, Shufesky WJ, et al. Mechanism of transfer of functional microRNAs between mouse dendritic cells via exosomes. Blood 2012;119(3):756–66. Available from: https://doi.org/10.1182/blood-2011-02-338004.
[10] Tkach M, Kowal J, Zucchetti AE, et al. Qualitative differences in T-cell activation by dendritic cell-derived extracellular vesicle subtypes. EMBO J 2017;36(20):3012–28. Available from: https://doi.org/10.15252/embj.201696003.
[11] Wang Y, Zheng F, Gao G, et al. MiR-548a-3p regulates inflammatory response via TLR4/NF-κB signaling pathway in rheumatoid arthritis. J Cell Biochem 2018;120 (2):1133–40. Available from: https://doi.org/10.1002/jcb.26659.
[12] Shivdasani RA. MicroRNAs: regulators of gene expression and cell differentiation. Blood 2006;108(12):3646–53. Available from: https://doi.org/10.1182/blood-2006-01-030015.
[13] Pomatto MAC, Bussolati B, D'Antico S, et al. Improved loading of plasma-derived extracellular vesicles to encapsulate antitumor miRNAs. Mol Ther Methods Clin Dev 2019;13:133–44. Available from: https://doi.org/10.1016/j.omtm.2019.01.001.
[14] Benmoussa A, Lee CH, Laffont B, et al. Commercial dairy cow milk microRNAs resist digestion under simulated gastrointestinal tract conditions. J Nutr 2016; 146(11):2206–15. Available from: https://doi.org/10.3945/jn.116.237651.
[15] Lamichhane TN, Jay SM. Production of extracellular vesicles loaded with therapeutic cargo. Methods Mol Biol 2018;1831:37–47. Available from: https://doi.org/10.1007/978-1-4939-8661-3_4.
[16] Kahn S, Liao Y, Du X, Xu W, Li J, Lönnerdal B. Exosomal microRNAs in milk from mothers delivering preterm infants survive in vitro digestion and are taken up by human intestinal cells. Mol Nutr Food Res 2018;62(11):e1701050. Available from: https://doi.org/10.1002/mnfr.201701050.
[17] Ma J, Wang C, Long K, et al. Exosomal microRNAs in giant panda (*Ailuropoda melanoleuca*) breast milk: potential maternal regulators for the development of newborn cubs. Sci Rep 2017;7(1):3507. Available from: https://doi.org/10.1038/s41598-017-03707-8.
[18] Torregrosa Paredes P, Gutzeit C, Johansson S, et al. Differences in exosome populations in human breast milk in relation to allergic sensitization and lifestyle. Allergy 2014;69(4):463–71. Available from: https://doi.org/10.1111/all.12357.

CHAPTER NINE

Human milk in the neonatal intensive care unit: dogma meets science

"A central lesson of science is that to understand complex issues (or even simple ones), we must try to free our minds of dogma and to guarantee the freedom to publish, to contradict and to experiment. Arguments from authority are unacceptable."

Carl Sagan

Key points

1. There are numerous benefits human milk incurs for preterm infants, but there are still many deficiencies that need to be addressed.
2. Providing appropriate energy, high-quality protein, lipids, and minerals specifically tailored toward individual needs are major challenges.
3. New technologies such as multiomics and artificial intelligence provide the potential for meeting these challenges.

Introduction

Preterm infants have very different nutritional needs when compared to term infants. Current neonatal intensive care unit (NICU) care entails providing energy on a per kilogram basis to preterm infants that is in the range of 120 kcal/kg per day. Protein needs are about 4–5 times on a per kilogram basis of a human adult. In the past decades, with the increased survival of extremely low birth weight infants, even higher energy intakes of 125–130 kcal/kg per day and protein intakes of 3.5–4 g/kg per day appear to be necessary to produce normal growth in extremely low birthweight (ELBW) infants [1]. To take this into

The Science of Human Milk.
DOI: https://doi.org/10.1016/B978-0-323-99913-7.00011-3

perspective, an average Tour de France cyclist requires up to 8000 calories per day. If this cyclist weighs 70 kg, on a per kilogram basis, he takes 114 calories/kg per day to meet this requirement, he takes in less than the ELBW requires for growth on a per kilogram basis.

Likewise, if one considers the average protein in a typical hamburger as 20 g, a typical 70 kg adult taking this amount of protein on a per kilogram basis would need to eat 14 hamburgers a day to meet the equivalent of the ELBW protein requirement, approximately 4 g/kg per day.

The preterm infant has calcium and phosphorus requirements that far exceed that of term infants [2]. Requirements for meeting demands incurred by metabolic synthetic pathway deficiencies include a supply of certain nutrients such as long-chain fatty acids that are readily synthesized in term infants if adequate essential fatty acid precursors are provided. These synthetic pathways are often deficient in preterm infants [3].

Human milk provided by the baby's own mother or banked donor milk has become the feeding of choice in the United States for a very low birth weight infant. Guidelines provided by the American Academy of Pediatrics in 2012 recommend that only human milk should be fed to these infants as seen in Fig. 9.1 [4].

As seen in the figure and quote below, there is a strong recommendation by the American Academy of Pediatrics for the use of human milk in these infants.

This is a recommendation being followed by most NICUs in the United States. However, to meet the requirements that provide adequate

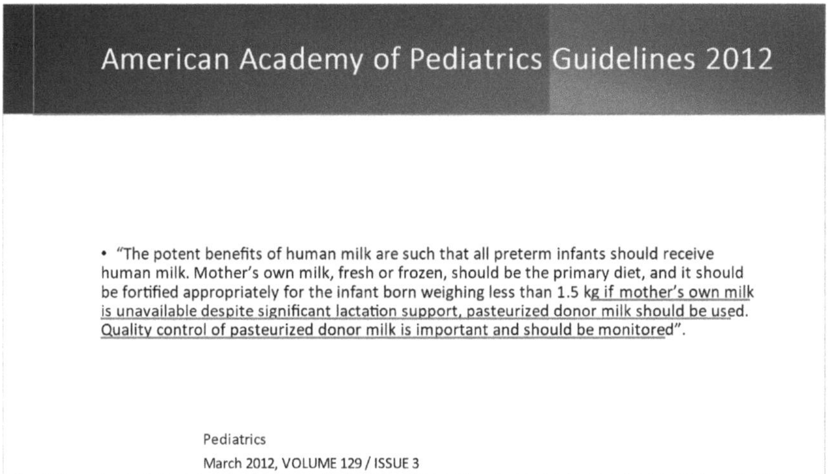

Figure 9.1 American Academy of Pediatrics statement on guidelines for providing human milk to preterms.

calories, protein, and other nutrients, large volumes of up to 300 mL/kg per day would need to be provided to meet the needs, which is unrealistic for most ELBW infants who are also sensitive to high fluid intakes because of fetal cardiovascular channels such as the patent ductus arteriosus and the propensity of the preterm infant lung to develop fluid overload and dispose to bronchopulmonary dysplasia. Thus the institution of the above guidelines, if strictly interpreted that only human milk (baby's own mother or donor) should be used, and if the fluids were restricted, this would provide insufficient nutrients for many of these very high-risk infants.

Donor milk was already being used extensively prior to the publication of these guidelines in 2012. The benefits of feeding premature infants their own mother's milk (MOM) are based on sound aspects of human physiology. However, the benefits of pooled, banked, donor milk are not supported by the same level of evidence as that supporting the use of the baby's own MOM, and the lack of many of the bioactive components in processed donor milk incurs a lack of biologic plausibility for having the same potent benefits. To consider that donor human milk as comparable in composition and health benefits is problematic.

Overinterpretation of the guidelines as suggesting equivalence between donor and baby's own MOM can be problematic. A study by Parker et al. compared the consumption of mothers' own milk by infants born extremely preterm following implementation of a donor milk program and after [5]. The study showed decreased use of mother's own milk after implementation of this program, which may be due to the misconception that the two types of milk are equivalent, and that donor milk provides the same benefits, which from the compositional perspective, it does not.

It is important to note that in the pasteurization process of donor milk, there is removal of numerous immune and bioactive components [6]. These include live cells such as leukocytes and macrophages, various enzymes such as lipase and alkaline phosphatase, and potentially beneficial microbes. Various immunoglobulin concentrations including IgA are also decreased. Donor milk from this perspective has a composition that lacks many of the bioactive factors found in babies own MOM. As with commercial formulas, donor milk lacks many properties that appear to be highly protective for preterm infants.

Furthermore, in addition to deficiencies in macronutrients and bioactive components, neither donor milk nor babies own MOM alone provides appropriate micronutrients to meet the requirements of these rapidly growing preterm infants. Baby's own MOM composition is dynamic and changes with the environmental conditions and thus needs of the infant [7].

Unfortunately, dogma has developed pertaining to donor milk which is not based and rigorous science. Such dogma may lead to problems such as undernutrition, growth failure, and deficiencies of certain individual nutrients The underlying reasons include the impact of pasteurization on immune components of breast milk and pooling of leftover milk donated from mothers who are already feeding their own infants, and the impact of lower direct mother–infant interactions.

Next, we more specifically address several questions as issues relating to these differences

What is the composition of donor human milk versus baby's own mother's milk?

MOM provided directly from the breast is the gold standard as all its biologically active components are preserved. Preterm infants require tube feeding and are either fed fresh expressed or frozen MOM, pasteurized DM, or formulas. Milk storage and processing affect bioactive components of MOM and DM. These differences include cellular components, immune factors such as lactoferrin, IgA and IgG, macronutrients, micronutrients, and enzymes.

Components of fresh MOM that are affected by standard Holder pasteurization include immunoglobulins, enzymes such as lipase, alkaline phosphatase and amylase, growth factors, and hormones [8,9].

Weight gain and growth with exclusive donor milk compared to formula milk also is of some concern. The following table is derived from a meta-analysis from Quigley and Embleton [10] that suggests an inferiority in growth in preterm infants fed with donor breast milk versus formula.

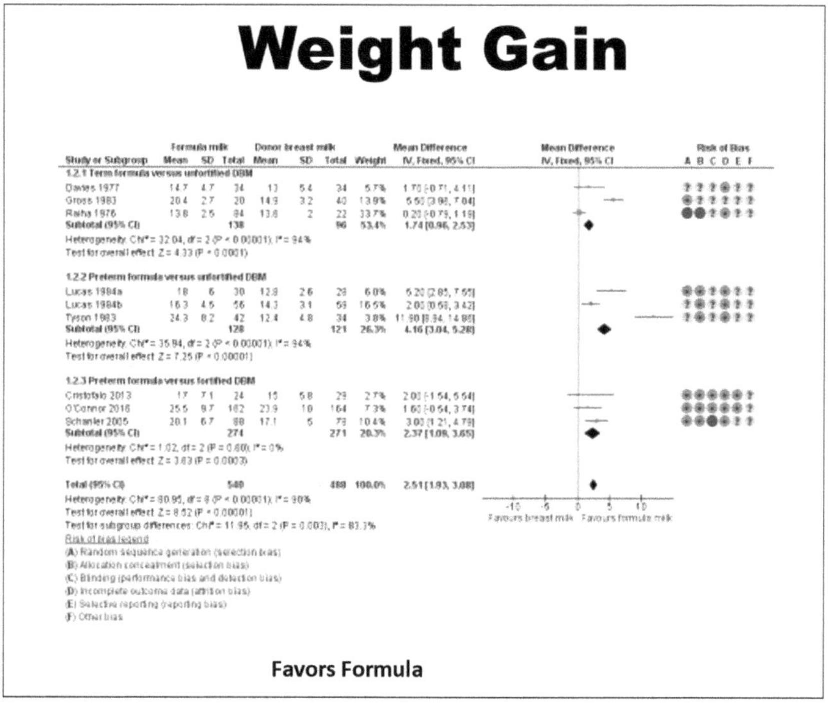

There are several concerns related to lipid metabolism and bioactive factors that are lost after pasteurization. Several enzymes are inactivated with the pasteurization process. Bile salt-stimulated lipase is inactivated, and this has implications in the efficiency of lipid digestion in the intestinal lumen (Fig. 9.2).

Furthermore, alkaline phosphatase is also inactivated, and this has implications in terms of mitigating the potency of lipopolysaccharide and its subsequent effect on inflammation (Fig. 9.3).

There are also issues with calcium and phosphorus in human milk. Critically ill infants in the NICU often receive low quantities of calcium and phosphorus. They commonly receive steroids as well as diuretics that can negatively affect bone development, thus are at high risk for fractures.

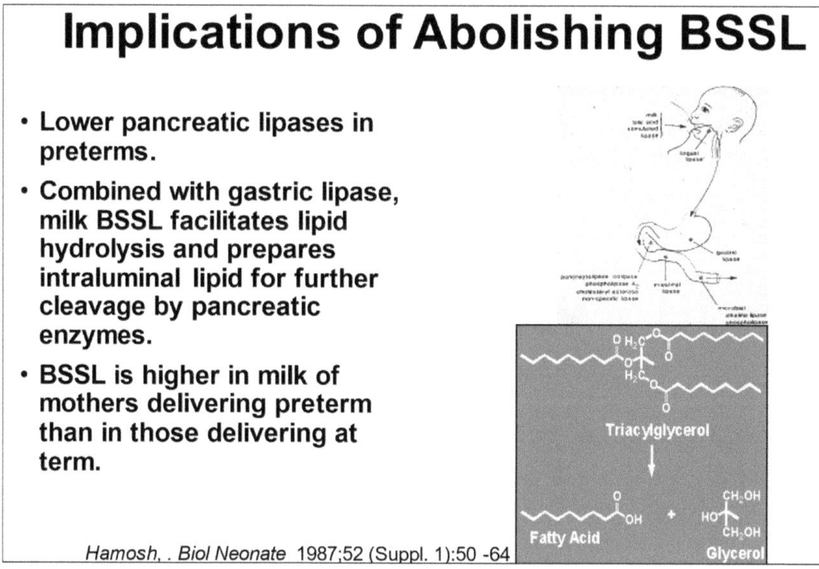

Figure 9.2 Bile salt-stimulated lipase is lost with pasteurization.

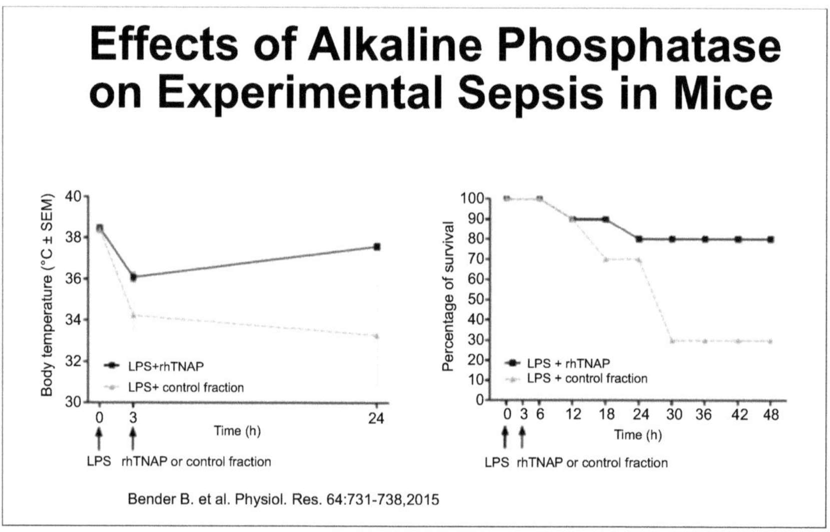

Figure 9.3 Effects of alkaline phosphatase on experimental sepsis.

Fig. 9.4 at the left shows a chest radiograph of an infant with severe osteopenia where the ribs are barely visible. This is partially due to mineral deficiency (calcium and phosphorus) and the requirements and content of calcium and phosphorus in various types of intakes [11].

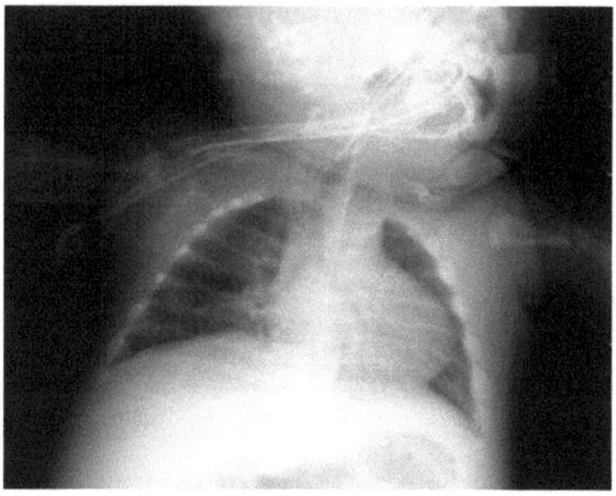

Figure 9.4 Osteopenia in a preterm infant. Please note the thin, gracile ribs. *From So K-W, Ng P-C. Treatment and prevention of neonatal osteopenia. Curr Paediatr 2005;15 (2):106–13. https://doi.org/10.1016/j.cupe.2004.12.011.*

Human milk microbes: are they pathogens, commensals, or just bystanders

Human milk microbes: Are they pathogens, commensals, or just bystanders? There has been a long-term concern about feeding preterm infants with their underdeveloped immune systems human milk that may be colonized with microbes. In fact, it has been common practice to culture human milk samples and discard them if microbes were found in any quantities considered "significant." Commercial formulas are sterilized and do not contain microbes and donor human milk is pasteurized and should contain few if any microbes. Recently, it has become clear that human milk contains a wide variety of taxa of microbes (perhaps around 700), many of which do not originate from the skin of the mother or infant or the infant's mouth [12–14]. Many of these microbes also originate from the mother's gastrointestinal tract. This is especially relevant because alterations of the maternal microbiota by diet or microbial therapy could be used to promote a breast milk microbial ecology that is most conducive to infant health. The exact mechanism of transfer of maternal intestinal microbes to the mothers' breast remains poorly understood but it has been speculated that the intestinal tract during pregnancy is more permeable than during the nonpregnant state and that microbial translocation may occur readily through this highly permeable intestine. Another mechanism involves the transfer of microbes via dendritic cells which underlie the intestinal mucosa and can send their appendages into the lumen of the intestine and subsequently enter the bloodstream and present these cells to the mothers' breast [15].

The microbial taxa from an individual MOM vary only slightly from week to week, but the microbes from an individual mother differ markedly from other mothers [16].

Whether this milk microbiome is specific for a certain mother–infant dyad and that this specificity could confer benefits to the infant through an enteromammary system described previously in Chapter 4.

Along the same thinking is the question of whether the technique of "kangaroo care" or "mother skin to skin care" may also offer benefits whereby the mother and infant are exposed to each other's microbes, and the mother acquires the infants' microbiota, which may offer specific benefits to the infant immune system. It is common practice to add fortifiers to either preterm baby's own MOM or donor milk. The effect of these on the infant intestinal microbiome is poorly understood.

Does donor human milk (HM) prevent necrotizing enterocolitis?

Necrotizing enterocolitis (NEC) and late-onset sepsis (LOS) are major problems in NICU care and account for significant suffering as well as increased costs. Studies that relate to prevention of these diseases in human preterm infants have been hampered by lack of good definitions of these entities. NEC, for example, is a simple single pathophysiologic pathway disease and thus preventative strategies may not target the most common pathway [17,18]. For example, when one considers using nutritional immune-related strategy for prevention of ischemic bowel disease (often termed "NEC") seen in infants with certain congenital heart problems, this does not have biological plausibility. On the other hand, the more classical forms of "NEC" seen in human preterms do appear to have a major immune-inflammatory mechanism as a component of its pathophysiology. Human milk microbes as well as immune components are thus likely to play a biologically plausible role in prevention of the more classic forms of NEC. The putative protective effects of MOM over DM and formula against NEC and LOS have been the subject of several studies, but most have been underpowered, and the individual studies have not shown a significant benefit of DM over formulas. Systematic reviews suggest that this may be the case [19], but are far from convincing. Studies need to be taken in the context of when these studies were done and how "NEC" and sepsis have been defined. Newer studies are ongoing and preliminary results suggest that DM may not have major benefits over formula in terms of NEC or LOS.

Are there differences in growth and development of preterm infants with DM versus MOM versus formula?

There are major compositional differences in macro- and micronutrients when comparing MOM, DM, and preterm formulas. The energy and protein content of human milk when provided to very low birth weight and extremely low birth weight infant do not meet their needs for growth and may in fact be detrimental from neurodevelopmental perspective unless they are fortified. Strategies have been developed for transitioning from parenteral nutrition to human milk, but these have often been met with at least transient deterioration in growth. Using fortifiers has been helpful, but specific strategies on when and how to institute their incorporation most safely are still being evaluated [20].

In summary donor milk is not a panacea. Why?

- Many of the active components including live microbial components, cells, and enzymes that may provide benefits are lost in the processing of donor milk.
- The mother provides personalization and precision. Can we ever mimic this?

References

[1] Denne SC. Protein and energy requirements in preterm infants. Semin Neonatol 2001;6(5):377–82. Available from: https://doi.org/10.1053/siny.2001.0059.
[2] Mihatsch W, Thome U, Saenz de Pipaon M. Update on calcium and phosphorus requirements of preterm infants and recommendations for enteral mineral intake. Nutrients 2021;13(5). Available from: https://doi.org/10.3390/nu13051470.
[3] Lapillonne A, Groh-Wargo S, Gonzalez CH, Uauy R. Lipid needs of preterm infants: updated recommendations. J Pediatr 2013;162(3 Suppl):S37–47. Available from: https://doi.org/10.1016/j.jpeds.2012.11.052.

[4] SECTION ON BREASTFEEDING, Eidelman AI, Schanler RJ, et al. Breastfeeding and the use of human milk. Pediatrics 2012;129(3):e827–41. Available from: https://doi.org/10.1542/peds.2011-3552.
[5] Parker LA, Cacho N, Engelmann C, et al. Consumption of mother's own milk by infants born extremely preterm following implementation of a donor human milk program: a retrospective cohort study. J Pediatr 2019;211:33–8. Available from: https://doi.org/10.1016/j.jpeds.2019.03.040.
[6] Wesolowska A, Sinkiewicz-Darol E, Barbarska O, Bernatowicz-Lojko U, Borszewska-Kornacka MK, van Goudoever JB. Innovative techniques of processing human milk to preserve key components. Nutrients 2019;11(5):1169. Available from: https://doi.org/10.3390/nu11051169.
[7] Meier PP. More evidence: mothers' own milk is personalized medicine for very low birthweight infants. Cell Rep Med 2022;3(8):100710. Available from: https://doi.org/10.1016/j.xcrm.2022.100710.
[8] Peila C, Moro GE, Bertino E, et al. The effect of holder pasteurization on nutrients and biologically-active components in donor human milk: a review. Nutrients 2016;8(8):477. Available from: https://doi.org/10.3390/nu8080477.
[9] Binte Abu Bakar SY, Salim M, Clulow AJ, Nicholas KR, Boyd BJ. Human milk composition and the effects of pasteurisation on the activity of its components. Trends Food Sci Technol 2021;111:166–74. Available from: https://doi.org/10.1016/j.tifs.2021.02.055.
[10] Quigley M, Embleton ND, McGuire W. Formula versus donor breast milk for feeding preterm or low birth weight infants. Cochrane Database Syst Rev 2014;4:CD002971. Available from: https://doi.org/10.1002/14651858.CD002971.pub3.
[11] So K-W, Ng P-C. Treatment and prevention of neonatal osteopenia. Curr Paediatr 2005;15(2):106–13. Available from: https://doi.org/10.1016/j.cupe.2004.12.011.
[12] Fernández L, Pannaraj PS, Rautava S, Rodríguez JM. The microbiota of the human mammary ecosystem. Front Cell Infect Microbiol 2020;10:586667. Available from: https://doi.org/10.3389/fcimb.2020.586667.
[13] Lyons KE, Ryan CA, Dempsey EM, Ross RP, Stanton C. Breast milk, a source of beneficial microbes and associated benefits for infant health. Nutrients 2020;12(4):1039. Available from: https://doi.org/10.3390/nu12041039.
[14] Fernández L, Rodríguez JM. Human milk microbiota: origin and potential uses. Nestle Nutr Inst Workshop Ser 2020;94:75–85. Available from: https://doi.org/10.1159/000505031.
[15] Neu J. Gut microbiota, host gene expression, and cell traffic via milk. Nestle Nutr Inst Workshop Ser 2020;94:94–102. Available from: https://doi.org/10.1159/000504996.
[16] Hunt KM, Foster JA, Forney LJ, et al. Characterization of the diversity and temporal stability of bacterial communities in human milk. PLoS One 2011;6(6):e21313. Available from: https://doi.org/10.1371/journal.pone.0021313.
[17] Neu J, Modi N, Caplan M. Necrotizing enterocolitis comes in different forms: historical perspectives and defining the disease. Semin Fetal Neonatal Med 2018;23(6):370–3. Available from: https://doi.org/10.1016/j.siny.2018.07.004.
[18] Neu J. Necrotizing enterocolitis: the future. Neonatology 2020;117(2):240–4. Available from: https://doi.org/10.1159/000506866.
[19] Cacho NT, Parker LA, Neu J. Necrotizing enterocolitis and human milk feeding: a systematic review. Clin Perinatol 2017;44(1):49–67. Available from: https://doi.org/10.1016/j.clp.2016.11.009.
[20] Hair AB, Scottoline B, Good M. Dilemmas in human milk fortification. J Perinatol 2022;43:103–7. Available from: https://doi.org/10.1038/s41372-022-01502-6.

CHAPTER TEN

Medications and toxins in mother's milk

Key Points

1. Many medications taken by the mother can be readily passed into breast milk.
2. Many toxins lactating mothers are exposed to can also be readily passed into breast milk.
3. Most medications do not incur immediate detrimental effects on the fetus but should still be considered for their potential long-term effects such as antibiotic resistance in the infant as well as metabolic consequences.
4. Breastfeeding is not always contraindicated in mothers with illicit drug use and requires careful consideration of risks and/or benefits for the infant.

All things are poison, and nothing is without poison; the dosage alone makes it so a thing is not a poison.

Paracelsus

Introduction

In this chapter, we will review concerns for medications found in breast milk as well as environmental toxins that might cause harm by transference through breast milk. Mothers commonly are anxious and discouraged about breastfeeding their infants if they are taking medications. Evidence for medication use during lactation is limited. However, current evidence supports that very few medications are absolutely contraindicated. Furthermore, suitable medication alternatives exist. Some of the

The Science of Human Milk.
DOI: https://doi.org/10.1016/B978-0-323-99913-7.00001-0

classes of medications that warrant seeking alternatives include radioactive compounds, amphetamines, chemotherapeutic agents, and statins.

Here we will review some of the information pertaining to various drugs that are of concern and others that may be problematic under certain circumstances and some that do not present concerns but are often misunderstood. The focus is on ones commonly used that may be of concern during breastfeeding.

Antibiotics

Antibiotics are among the most commonly prescribed drugs in the United States [1]. Antibiotics are prescribed to postpartum women for a variety of reasons. These include cesarean sections, premature rupture of membranes, urinary tract infections, sexually transmitted diseases, vaginitis, mastitis, giardiasis, and respiratory tract infections. The use of antibiotics is essential and can prevent patients from experiencing important morbidity and/or mortality. Most antibiotics given to mothers are safe for their infants during breastfeeding [2]. Antibiotics that are considered safe if given directly to the infant are also considered safe for the infant of the lactating mother receiving the antibiotic. When assessing the tolerability of a particular antibiotic in a breast-feeding infant, it is important to initially determine whether the antibiotic is well tolerated when given directly to an infant. How the antibiotic modifies the intestinal flora of the infant and may cause subsequent dysbiosis, emergence of resistant strains and the effects on the developing immune system should also be considered, but such specific information is not readily available.

In the majority of instances antibiotics do not distribute into milk in sufficient concentrations to be of any clinical consequence in the breastfeeding infant [2]. The following are some considerations for specific antibiotics.

1. *Fluoroquinolone* antibiotics such as ciprofloxacin are often recommended during breastfeeding. The amount of fluoroquinolone a breast-fed infant could actually receive during nursing would appear to be negligible [3]. Despite the weakness of evidence for harm, the manufacturer of ciprofloxacin states "Ciprofloxacin is excreted in human milk. Because of the potential for serious adverse reactions in

infants nursing from mothers taking ciprofloxacin, a decision should be made whether to discontinue nursing or to discontinue the drug, taking into account the importance of the drug to the mother" [4]. These recommendations stem from concerns previously described in juvenile animals over the development of fluoroquinolone-induced arthropathies and cartilage erosion [5]. It is recommended by some authors that if a fluoroquinolone is the optimal antibiotic to treat a breast-feeding mother, the breast-feeding infant should continue breastfeeding without interruption [2].

2. *Sulfonamide* use in mothers who breastfeed their infants has been discouraged during the first month of life because of a theoretical risk of bilirubin displacement from albumin binding sites by sulfonamide, resulting in kernicterus. Although this possibility is posed in the literature, published reports of kernicterus developing in an infant solely exposed to a sulfonamide through breastfeeding do not exist [6]. An extensive systematic review of the use of sulfonamides near term and during breastfeeding found no adverse reactions in infants. The authors concluded that the use of sulfamethoxazole-trimethoprim during breastfeeding presents no risk of neonatal kernicterus [7].
3. *Macrolide and Azilide Antibiotics—erythromycin and azithromycin* are usually well tolerated when administered directly to neonates, infants, and children. Erythromycin-induced pyloric stenosis resulting from breastfeeding exposure in an infant aged 3 weeks has been reported [8]. In this report, the infant's stools turned an orange-red color upon starting the erythromycin but were guaiac-negative. Five days into the maternal erythromycin treatment, the infant developed persistent projectile vomiting. After the maternal erythromycin was discontinued, continued vomiting and weight loss prompted an upper GI evaluation which revealed hypertrophic pyloric stenosis that was corrected with surgery. Reviews of the association between macrolide antibiotic use during pregnancy and lactation and the development of hypertrophic pyloric stenosis show either no significant association or are inconclusive [9,10]. The AAP describes erythromycin as a medication that is concentrated in breast milk but is usually compatible with breastfeeding [11].
4. *Tetracyclines' effects* upon developing teeth and bones have been well characterized, particularly as they pertain to infants and children [12]. These adverse effects have been documented in children who receive the drug directly, especially after courses. There have been

recommendations against breastfeeding while mothers are receiving tetracycline or tetracycline analogs. However, clinical evidence is based on theoretical concerns rather than supported with clinical data. The AAP recommendation [13] is that tetracycline analogs are safe for use in lactating mothers and that breastfeeding should be continued in breastfeeding mothers who are receiving such therapy.

Pain medications

Nonsteroidal antiinflammatory drugs, such as ibuprofen have minimal transfer into milk and are good options for pain management for breastfeeding women. One theoretical concern relates to the use of such drugs in breastfeeding mothers of infants with ductal-dependent heart lesions. Such drugs have the capability of closing the ductus arteriosus with potentially devastating consequences. Opioids transfer into human milk and may sedate the infant. Thus opioids should be used for the shortest period and at the lowest doses that produce adequate analgesia. Codeine and tramadol pose risks because of high interpersonal variability of CYP2D6, an enzyme that catalyzes the metabolism of many clinically important drugs including antidepressants, neuroleptics, some antiarrhythmics, lipophilic β-adrenoceptor blockers, and opioids. With low CYPsD7 activity, high levels of the metabolite may develop and thus should be avoided.

Contraception

Contraceptive counseling should include discussion of the effectiveness of the chosen method as well as the likelihood of its effect on milk supply. Contraception is important for lengthening birth intervals and improving the health of women and children. Nonhormonal methods of contraception are the preferred choice for breastfeeding women. Hormonal methods, including progesterone-only or combined pills, implants, patches, and intrauterine devices, are all safe for breastfed infants but have varying effects on lactation [14].

Illicit drugs

According to the 2016 National Survey on Drug Use and Health (https://www.samhsa.gov/data/sites/default/files/NSDUH-DetTabs-

2016/NSDUH-DetTabs-2016.pdf), one in 10 people (28 million) over age 12 used an illicit drug in the past 30 days. Of pregnant women, 6.3% used illicit drugs, 8.3% reported alcohol use, and 4.3% reported binge drinking in the last month.

There are numerous benefits to breastfeeding which include child spacing, reduced sudden infant death syndrome, and infections. These benefits may be of particular importance to mothers struggling with substance abuse. Numerous organizations including the American Academy of Pediatrics, the Centers for Disease Control, and the American College of Obstetricians and Gynecologists have made statements related to drug use during breastfeeding (see Table 10.1).

Screening for drug use

Lactation cannot ethically be regulated or banned by a medical or government authority. Recommendations as to whether to breastfeed should be free from intrinsic or organizational biases. Alternate nutrition should be considered with maternal permission when substance abuse treatment and/or mental health stability are in significant jeopardy.

Ideally, breastfeeding is most successful in those who are abstinent from substance use, and/or part of a comprehensive substance use treatment program. If abstinence is not possible, harm reduction strategies such as compliance with provider visits, seeking an alternative childcare provider, and avoidance of cosleeping with the baby when using drugs should be employed.

Impact of specific substances on breastfeeding

Table 10.2 summarizes breastfeeding recommendations with specific drugs.

Next, we will discuss some aspects of specific drugs as they related to breastfeeding.

Methadone

The concentration of methadone in human milk is low and women on methadone maintenance should be encouraged to breastfeed regardless of

Table 10.1 Statements from organizations regarding breastfeeding and substance use.

Organization	Statement
American Academy of Pediatrics 2012	Maternal substance abuse is not a categorical contraindication to breastfeeding. Street drugs such as phencyclidine (PCP), cocaine, and cannabis can be detected in human milk, and their use by breastfeeding mothers is of concern, particularly regarding the infant's long-term neurobehavioral development, and thus are contraindicated.
American Academy of Pediatrics 2018	Although this (the 2012 recommendation) has been interpreted by some professional organizations to indicate that in the parent using marijuana, the choice to breastfeed is "contraindicated," this was not the intent of that statement. It is suggested instead that the mother be encouraged to breastfeed while, at the same time, it is strongly encouraged that she abstains completely from using marijuana as well as other drugs, alcohol, and tobacco.
Centers for Disease Control	Mothers should not breastfeed or feed expressed milk to their infants if they are using an illicit street drug, such as PCP or cocaine. (Exception: Narcotic-dependent mothers who are enrolled in a supervised medication-assisted treatment program and have a negative screening for HIV infection and other illicit drugs can breastfeed).
American College of Obstetricians and Gynecologists 2018	A woman with current or past history of substance use disorder should not be summarily excluded from or criminalized for nursing her infant. For women in well-supervised medication-assisted treatment programs, breastfeeding is encouraged. Obstetrician-gynecologists must have accurate information about the potential dangers of illicit substances and high concentrations of alcohol through breast milk so they can accurately advise patients about benefits and harms of breastfeeding.

Source: From Bartholomew, LM, Lee MJ. Substance use in the breastfeeding woman. Contemporary OB/GYN J 2019;64(9):1–3.

Table 10.2 Breastfeeding recommendations under conditions of specific drug use.

Substance	Breastfeeding recommendations
Methadone	Yes Once neonatal abstinence syndrome resolved, feed before dose or discard milk for 2–4 h after dose
Buprenorphine	Yes
Other prescription opiates	Short-term use: Yes Long-term use: No until transition to medication-assisted treatment or monitoring by pain specialist Codeine use: No
Heroin	Short-term use: Yes, but wait or discard milk for 24–48 h after use (may be impractical) Frequent use: No
Marijuana	Yes, but support cessation and no smoke exposure to infant
Methamphetamine	Occasional use: Yes, but wait or discard milk for 48–100 h after use (impractical) Frequent use: No
Cocaine	Occasional use: Yes, but wait or discard milk for 24 h after use (may be impractical) Frequent use: No
Benzodiazepines	Occasional use: Yes, but wait or discard milk for 6–8 h after useFrequent or high dose use: No
Alcohol	Occasional use: Yes, but wait or discard for 90–120 min or 2 h per drink Frequent use: No
Tobacco	Yes, but support cessation, nicotine replacement, and avoid smoke exposure to infant

Source: Created from information found in Reece-Stremtan S, Marinelli K, The Academy of Breastfeeding Medicine. ABM Protocol #21. Guidelines for breastfeeding and substance use or substance use disorder, Revised 2015. Breastfeed Med 2015;10(3):135–41; Drugs & lactation database (Lactmed) NLM https://toxnet.nlm.nih.gov/ (accessed in June 2019); *Source*: From Bartholomew, LM, Lee MJ. Substance use in the breastfeeding woman. Contemporary OB/GYN J 2019;64(9):1–3.

methadone dose [15,16]. Neonatal abstinence syndrome (NAS) occurs in approximately 70% of neonates born to mothers prescribed methadone and may negatively influence latch and sucking. Because NAS is significantly reduced in neonates whose mothers breastfeed while on methadone maintenance, vigorous efforts to encourage and support mothers to breastfeed are needed. Once NAS is resolved, strategies to reduce long-term exposure to the neonate include feeding or pumping before daily dose and or waiting 2–4 hours after a dose [17].

Buprenorphine

Buprenorphine is a partial opioid agonist with few side effects. It is therapeutic for medical-assisted treatment of opioid substance abuse. Levels of buprenorphine are low in human milk and are not likely to cause negative effects in the infant. Breastfed infants whose mothers used buprenorphine have less severe NAS and have excellent rates of breastfeeding (76% breastfed at all and 66% were breastfeeding at 6 weeks postdelivery), as is the case with methadone [18]. Breastfeeding is thus encouraged for mothers using buprenorphine.

Codeine

Codeine use during breastfeeding is discouraged by the US Food and Drug Administration. Its metabolite is morphine and some infants with carriers of copies of the BYP2D6 gene may be ultra-rapid metabolizers, and susceptible to respiratory arrest with high levels of this drug in the bloodstream. This gene variant is more common in particular ethnic groups (28% in North Africans, 10% in Caucasians, and 3% in African Americans) [19].

Heroin

Heroin (diacetylmorphine) is metabolized into morphine, which has an elimination half-life of 2–3 hours. Heroin has an elimination half-life of 15–30 minutes, is rapidly excreted into breast milk, and causes dependence in the infant. Milk should be pumped and discarded for 24–48 h after use, but this may be impractical [20]. Mothers using heroin who cannot enter treatment or manage harm reduction strategies should be counseled not to breastfeed.

Cocaine

High concentrations of cocaine are found in breast milk from recreational users. The elimination half-life is 1 hour, and concentrations are eight times higher in breast milk than in the plasma of the user. Infants are extremely sensitive to cocaine because of immature inactivation enzymes. Infants exposed to cocaine in breast milk exhibit numerous symptoms including seizures, vomiting, high-pitched crying, and respiratory distress. Infant deaths have been reported [20]. Mothers using cocaine who do not enter treatment or cannot manage harm reduction strategies should be counseled not to breastfeed.

Methamphetamine or MDMA

Amphetamine-type stimulants are concentrated in breast milk in higher amounts (3–8 times) than in the plasma of the user. The elimination half-life is 6–12 hours. Infants breastfed by amphetamine users experience numerous symptoms and deaths from cardiopulmonary failure have been reported. Mothers using methamphetamine or MDMA who do not enter treatment or cannot manage harm reduction strategies should be counseled not to breastfeed.

Marijuana

The psychoactive compound in marijuana, delta-9-tetrahydrocannabinol, accumulates in breast milk depending on chronicity of use. The elimination half-life is 20–36 hours and up to 4 days in chronic users. As a result, the use of marijuana is discouraged during breastfeeding [20].

Alcohol

Alcohol consumption in breastfeeding women is not uncommon. Alcohol interferes with the milk ejection reflex and may reduce production. Human milk levels are similar to maternal blood alcohol levels. Long-term effects of alcohol exposure in breast milk are unknown. According to some experts, alcohol intake should be limited to only 8 oz of wine or two beers, and one has to wait for 2 hours after drinking to resume breastfeeding.

Tobacco

Nicotine is concentrated in breast milk in concentrations up to three times maternal plasma. However, there is no evidence that nicotine in breast milk represents a health risk to the infant. The benefits of breastfeeding are thought to outweigh the risks. Tobacco smokers are encouraged to breastfeed while trying to quit and eliminate infant smoke exposure [20].

Environmental toxins in breast milk

Environmental toxins appear to be commonly found in breast milk in a study of a cohort of nursing mothers [21]. Despite the fact that breastfeeding is beneficial for infants, the effects of chemicals that may be toxins in breast milk should be taken seriously. Environmental pollutants, even if only found in trace quantities, may act together where coexposure may have significant impacts on their development and subsequent health.

Summary

While a large majority of medications taken by the lactating mother are safe for infants, it is to be remembered that part of this is due to lack of evidence, which does not mean that harm may still be incurred. Hence, if there is minimal or no need for the drug in the mom, it may be best to avoid it completely.

References

[1] McCaig LF, Hughes JM. Trends in antimicrobial drug prescribing among office-based physicians in the United States. JAMA 1995;273(3):214–19.
[2] de Sá Del Fiol F, Barberato-Filho S, de Cássia Bergamaschi C, Lopes LC, Gauthier TP. Antibiotics and breastfeeding. Chemotherapy 2016;61(3):134–43.
[3] O'Donnell JA, Gelone SP. Fluoroquinolones. Infect Dis Clin North Am 2000;14 (2):489–513.
[4] Arky R., Sifton D.W. Physicians' desk reference. Physicians Desk Reference Incorporated; 1994.
[5] Burkhardt JE, Hill MA, Carlton WW. Morphologic and biochemical changes in articular cartilages of immature beagle dogs dosed with difloxacin. Toxicol Pathol 1992;20(2):246–52.
[6] Esbjörner E, Järnerot G, Wranne L. Sulphasalazine and sulphapyridine serum levels in children to mothers treated with sulphasalazine during pregnancy and lactation. Acta Paediatr Scand 1987;76(1):137–42.
[7] Forna F, McConnell M, Kitabire FN, Homsy J, Brooks JT, Mermin J, et al. Systematic review of the safety of trimethoprim-sulfamethoxazole for prophylaxis in HIV-infected pregnant women: implications for resource-limited settings. AIDS Rev 2006;8(1):24–36.
[8] Stang H. Pyloric stenosis associated with erythromycin ingested through breastmilk. Minn Med 1986;69(11):669–70 82.
[9] Almaramhy HH, Al-Zalabani AH. The association of prenatal and postnatal macrolide exposure with subsequent development of infantile hypertrophic pyloric stenosis: a systematic review and meta-analysis. Italian J Pediatr 2019;45(1):20.
[10] Abdellatif M, Ghozy S, Kamel MG, Elawady SS, Ghorab MME, Attia AW, et al. Association between exposure to macrolides and the development of infantile hypertrophic pyloric stenosis: a systematic review and meta-analysis. Eur J Pediatr 2019;178(3):301–14.
[11] American Academy of pediatrics Committee on Drugs. The transfer of drugs and chemicals into human milk. Pediatrics. 1994;93:137–50.
[12] Grossman ER, Walchek A, Freedman H. Tetracyclines and permanent teeth: the relation between dose and tooth color. Pediatrics. 1971;47(3):567–70.
[13] American Academy of Pediatrics Committee on Drugs. Transfer of drugs and other chemicals into human milk. Pediatrics 1989;84(5):924–36.
[14] Pieh Holder KL. Contraception and breastfeeding. Clin Obstet Gynecol 2015;58 (4):928–35.

[15] Liu AJ, Nanan R. Methadone maintenance and breastfeeding in the neonatal period. Pediatrics 2008;121(4):869 author reply - 70.
[16] Ballard JL. Treatment of neonatal abstinence syndrome with breast milk containing methadone. J Perinat Neonatal Nurs 2002;15(4):76–85.
[17] Bogen DL, Whalen BL. Breastmilk feeding for mothers and infants with opioid exposure: what is best? Semfetal Neonatal Med 2019;24(2):95–104.
[18] O'Connor AB, Collett A, Alto WA, O'Brien LM. Breastfeeding rates and the relationship between breastfeeding and neonatal abstinence syndrome in women maintained on buprenorphine during pregnancy. J Midwifery Women's Health 2013;58 (4):383–8.
[19] Dean L, Kane M. Codeine therapy and CYP2D6 genotype. In: Pratt VM, Scott SA, Pirmohamed M, Esquivel B, Kattman BL, Malheiro AJ, editors. Medical genetics summaries. Bethesda (MD): National Center for Biotechnology Information (US); 2012.
[20] Anderson PO. Drugs of abuse during breastfeeding. Breastfeed Med 2018;13 (6):405–7.
[21] Rovira J, Martínez MÁ, Mari M, Cunha SC, Fernandes JO, Marmelo I, et al. Mixture of environmental pollutants in breast milk from a Spanish cohort of nursing mothers. Environ Int 2022;166:107375.

CHAPTER ELEVEN

The anatomy and physiology of lactation

"Imagine that the world had created a new "dream product" to feed and immunize everyone born on earth. Imagine also that it was available everywhere, required no storage or delivery, and helped mothers plan their families and reduce the risk of cancer. Then imagine that the world refused to use it."

Frank A. Oski

Key points

1. The anatomy and physiology of milk production in the breast are summarized.
2. Regulatory pathways for production of some of the major nutrients in milk are described.

The anatomy of female breast

Development

In the primitive mammal such as the monotreme platypus, milk is expressed directly from the ducts onto the fur, where the young lap it up. There are no nipples, and the mammary glands are functional in both sexes. In the kangaroo, the tiny newborn "joey" sucks the nipple, which expands in its mouth and attaches the young to the mother's body and nurses at will. In some mammals such as horses, and whales, the mammary glands are in the groin; in primates, they are on the chest. Most small mammals have several pairs spread along the ventral surface.

Mammary glands are a modification of sweat glands. They first appear in the embryo as clumps of cells proliferating from a longitudinal ridge of

The Science of Human Milk.
DOI: https://doi.org/10.1016/B978-0-323-99913-7.00008-3

ectoderm (the outermost of the three germ layers of the embryo) along the so-called milk line. These clumps ultimately become breasts and vary with each mammalian species according to the size of its litter. In the human normally only one develops on each side of the chest. As a girl approaches her teenage years, breast development begins. The ovaries produce and release estrogen and this is associated with the collection of fat and the connective tissue of the breast. This causes the breasts to enlarge. The duct system also begins to grow. With the beginning of ovulation and menstruation, the breast matures accordingly with the formation of the secretory glands at the ends of the milk ducts. Many glands and lobules continue to grow and mature.

Each month with the menstrual cycle, estrogen is produced by the ovaries in the first half of the menstrual cycle. This stimulates the growth of the milk ducts in the breast. In the second half of the menstruation cycle, the hormone progesterone increases and stimulates the formation of the milk glands. These fluctuations in estrogen and progesterone are believed to be responsible for the cyclical changes that occur in the breasts during the menstrual cycle that are also accompanied by swelling pain and soreness just before menstruation.

Breast anatomy

The anatomy of the breast is seen below in Fig. 11.1. The mammary glands inside the female breast are responsible for milk production. The average woman has 15–20 lobes within each breast. Each lobe is subdivided into many smaller lobules, separated by broad fibrous suspensory bands, which connect the skin with the fascia, or sheet of connective tissue. Each lobe is drained by a separate excretory milk duct. These converge beneath the nipple, where they widen into milk reservoirs, before narrowing again to emerge as pinpoint openings at the summit of the nipple. Circular and radiating muscles in the areola, a circular disk of roughened pigmented skin surrounding the nipple, cause the nipple to become firm and erect upon suckling. The areola also contains sebaceous glands to provide lubrication for the nipple during nursing.

Several areas within the mammary glands help to produce, store, and excrete milk:

- *Alveoli*—Milk is produced and stored within these small sacs. These sacs are clustered together to form lobes.

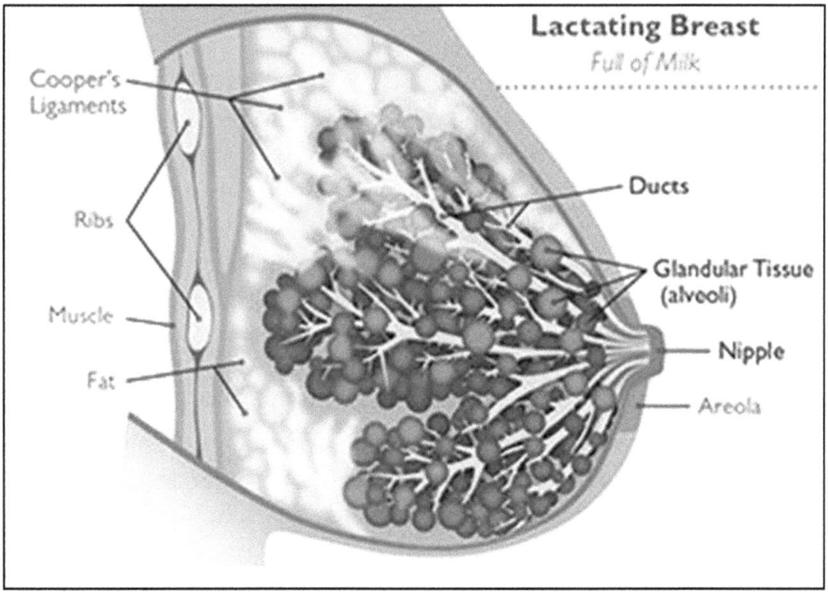

Figure 11.1 Anatomy of the breast.

- *Lactiferous ducts* (milk ducts)—The lobes are connected to milk ducts, and on average you can have 14–18 ducts per breast. The milk ducts carry milk to the nipple.
- *Nipple*—Milk is released through many tiny pores in the nipple.
- *Areola*—This pigmented area around your nipple helps stimulate milk flow when a baby latches on and takes the whole area into their mouth.

Lactation

Production of milk, or lactation, in female mammals occurs after giving birth. Hormones triggered by delivery of the placenta and by nursing stimulate milk production and secretion. The first milk that is produced is called colostrum it has more proteins, minerals, and antibodies but less fats than mature milk that develops later. Of interest is that if an infant is born prematurely, milk production still ensues. This is suggestive that milk production is held back during pregnancy. The mechanism by which this occurs is likely to involve several opposing actions of hormones such as estrogen, progesterone, and prolactin. Estrogens and progesterone during pregnancy appear to block the release of prolactin from the pituitary gland and thus the mammary gland is unresponsive. At the end of pregnancy with the expulsion of the placenta, some of these hormones are lost and the secretion of prolactin by the pituitary gland favors lactation.

Continued lactation is supported by the stimulus of nursing. This promotes the secretion of prolactin and triggers the release of another hormone from the pituitary gland called oxytocin. Oxytocin promotes the contraction of muscle cells around the alveoli and the breast, which promotes the expulsion of milk (Fig. 11.2).

Of interest is that a baby suckling at one breast may cause an increase in milk flow from both and this is likely due to this hormonal stimulation.

For continued breast milk production, a positive feedback loop occurs. Suckling at the breast stimulates the hypothalamus of the brain to relay a signal to the posterior pituitary gland, which secretes oxytocin. Oxytocin is a hormone (also used in the onset of labor) that sends a chemical signal to the cells surrounding the milk ducts, telling them to contract and release milk from the breast. Milk continues being produced and released until the baby stops suckling.

Mammary growth begins during pregnancy, and small amounts of milk are formed. However, milk secretion largely starts after delivery. Nursing is stimulatory and supports continued lactation. In two ways: it promotes the secretion of prolactin (and possibly other pituitary hormones of value in milk formation), and it triggers the release of yet another hormone from the pituitary gland—oxytocin, which causes the contraction of special muscles around the alveoli in the breast and ensures the expulsion of milk. This is the way that a baby's sucking at one breast may cause an increase in milk flow from both, so that milk may drip from the unsuckled nipple. About 30 seconds elapse between the beginning of active suckling and the initiation of milk flow.

The various detailed aspects of milk component production such as proteins, lactose, and lipids as well as the bioactive components are beyond the scope of this chapter and will briefly be summarized. It is important to note that in addition to the endogenous synthesis of various milk components, the composition of the mother's milk is also partially dependent on her diet.

Although the molecular details of milk synthesis in humans are not all fully understood, there has been progress in elucidating the pathways and their regulation of two of the major components, casein, the major milk protein, and lactose, the primary and milk-specific carbohydrate. We will focus on these two components.

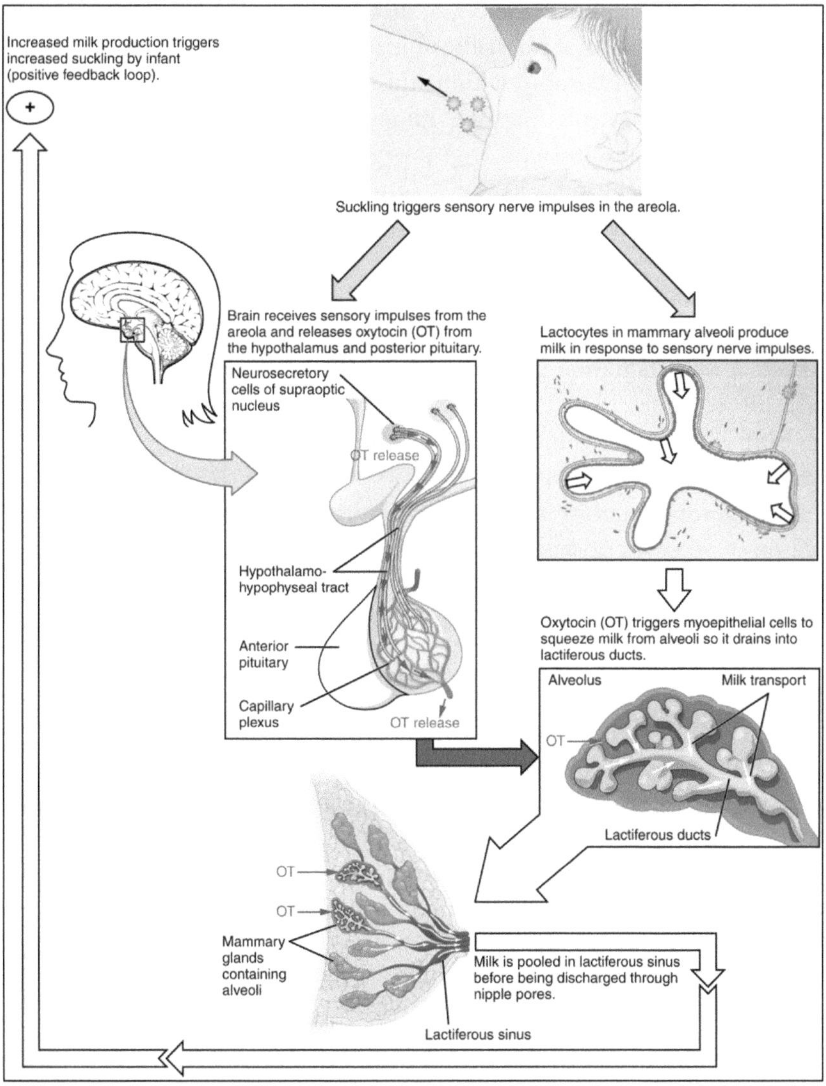

Figure 11.2 Oxytocin promotion of muscle cell contraction around the alveoli promoting the expulsion of milk.

Casein

Casein is a mixture of at least three different subtypes of the protein. Each is a polypeptide of 150–200 amino acids. These subtypes are synthesized in the endoplasmic reticulum of the epithelial glandular cells of the breast and undergo posttranslational modifications such as phosphorylation which allows them to form stable micelles which include calcium ions. These micelles are transported from the Golgi complex to acinus where they undergo exocytosis into the lumen. Transcription of the β-casein gene is stimulated by prolactin.

Lactose, the main disaccharide is composed of one molecule of galactose lactose and one molecule of glucose. It is synthesized in the Golgi complex of the epithelial cells of the breast alveoli. The irreversible formation of lactose is catalyzed by the lactose synthesizing system, composed of a 1:1 complex of β4-galactosyl transferase and lactalbumin. The expression of lactalbumin is elevated prior to parturition, whereas most of the other enzymes in the pathway increase in the first 4 days following the birth of the newborn. Milk is secreted from the epithelial cells into the duct as secretory granules.

Progesterone inhibits and prolactin stimulates milk synthesis. These hormones are two critical points of control. In addition, many other hormones, including glucocorticoids, growth hormone, thyroxin, and insulin, are required for the maintenance of normal lactogenesis.

Weaning and cessation of lactation

The cessation of breastfeeding is highly variable. It often varies between countries and socioeconomic status adds mother's ability to maintain breastfeeding while also working may play a role. In some countries and ethnicities, women of higher economic status tend to transition to artificial feeding early, whereas others may wait longer. Professional societies and pediatricians in general have concluded that there is no nutritional superiority or psychological benefit that results from the introduction of solid foods into the infant diet earlier than around 3–4 months and most full-term infants can thrive on a diet consisting exclusively of mother's milk.

With reduced demand of the infant, lactation slowly did declines and stops. Menstruation usually resumes within 6–8 weeks in women who are not breastfeeding. Although ovulation is less frequent during lactation, it does occasionally occur and thus breastfeeding should not be used as a method of contraception.

Summary

In this chapter various aspects of breast anatomy and physiology during lactation have been summarized. Lactation is a hormonally mediated process and is dependent on parturition of the infant. The composition of milk depends partially on the health and diet of the mother. Professional societies support exclusive breastfeeding in healthy infants up until at least 6 months after birth.

CHAPTER TWELVE

Benefits to the mother from breastfeeding

"Being a mom has made me so tired. And so happy."

Tina Fey

Key points

1. Despite challenges incurred to the mother by breastfeeding there are several benefits.
2. Among these benefits are weight loss, reduction of heart disease, cancer and type 2 diabetes risk and improved sleep.

Benefits shortly after birth

Oxytocin release

When a newly delivered mother has skin-to-skin contact with her newborn, oxytocin, "the love hormone" or "the cuddle chemical," is released and is particularly stimulated for release during the act of suckling [1,2].

As illustrated in Figs. 12.1 and 12.2, above (from https://www.infant-journal.co.uk/pdf/inf_054_ers.pdf) pathways related to oxytocin include activation by:

1. Sucking of the mother's nipple, in which the sensory nerves originate in the breast.
2. Sucking in the infant, in which the sensory nerves originate in the infant's oral mucosa.
3. The presence of food in the gastrointestinal tract, affecting vagal sensory nerves.

The Science of Human Milk.
DOI: https://doi.org/10.1016/B978-0-323-99913-7.00013-7

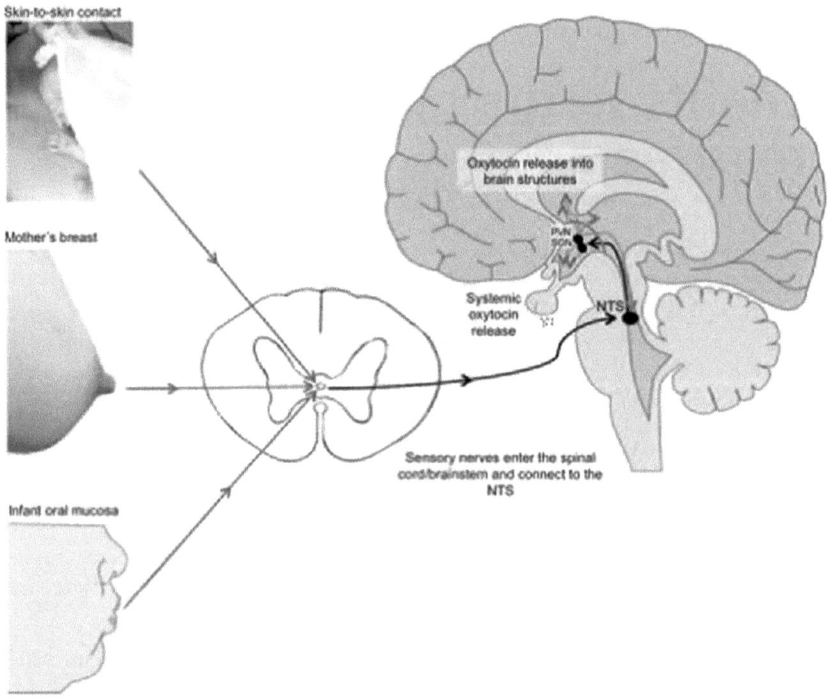

Figure 12.1 Sensory nerves release oxytocin during mother-infant interaction. While various sensory nerves can initiate this pathway, the nucleus tractus solitarius (NTS) acts as a common relay station for sensory input to the oxytocin-producing paraventricular (PVN) and supraoptic (SON) nuclei.

4. Skin-to-skin contact in both mothers and infants, in which the sensory nerves that originate in the skin respond to warmth, touch, stroking, and light pressure.

In addition, early latching and breastfeeding help the uterus contract back to its prepregnancy size, encourage expulsion of the placenta, and protect from the loss of blood [3]. In the days after a baby is born, the oxytocin-produced body during breastfeeding also helps prevent further blood loss, thus decreasing the likelihood of developing iron deficiency anemia [4]. Exclusive breastfeeding also largely halts menstruation. This is considered by some as a possible method of contraception but is not recommended because of its frequent ineffectiveness [5–7].

Oxytocin also has an antidepressant effect [8]. Mothers with higher levels of oxytocin have demonstrated fewer anxiety and depression

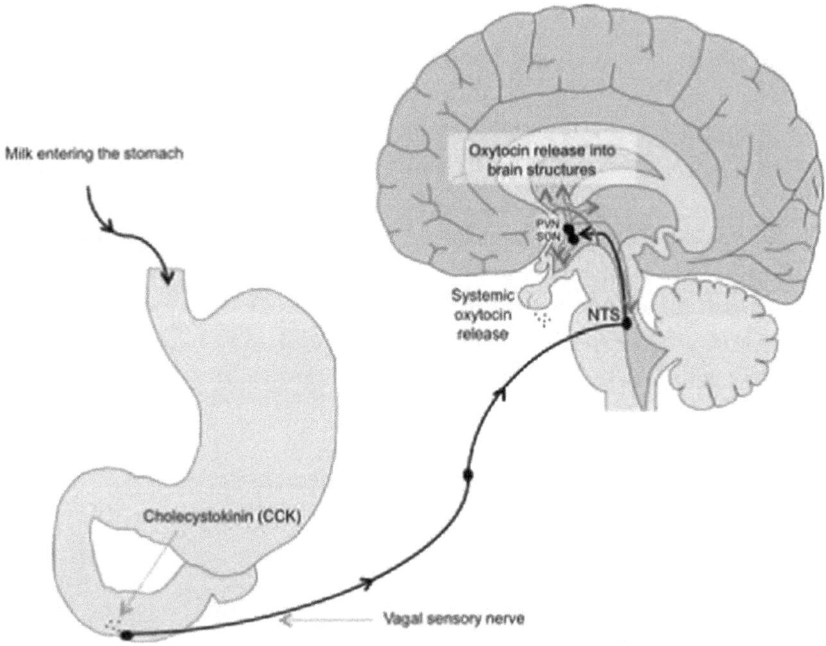

Figure 12.2 Food in the stomach triggers oxytocin release via the gut hormone cholecystokinin (CCK).

symptoms. In fact, oxytocin release during breastfeeding promotes stress reduction, decreases blood pressure, and increases the pain threshold [9].

Weight loss

About 500 additional calories are used by lactating mothers per day [10]. These extra calories burned with breastfeeding may help mothers shed some of the weight and fat mass they may have gained during pregnancy.

Heart disease, cancer, and type 2 diabetes

Breastfeeding lowers lifelong risk of heart disease and type 2 diabetes [11,12].

In a study that evaluated 52,731 women it was shown a 14% reduced likelihood of diabetes per year of breastfeeding. Compared with nulliparous women, childbearing women who do not breastfeed have about 50% increased risk of type 2 diabetes in later life [13].

Every extra month of breastfeeding reduces the risk of several cancers, including those of the breast, ovaries, and uterus [14,15].

Sleep

Breastfeeding may help the mother sleep more and better than if the infant is formula fed. In a study evaluating parents of infants who were breastfed versus formula fed, those parents of infants who were breastfed slept 45 minutes more than parents of infants given formula [16]. Parents given formula at night also self-reported more sleep disturbance than parents of infants who were exclusively breastfed at night. The conclusion from this study is that parents who supplement their infant feeding with formula and may have the impression that they will get more sleep should be encouraged to continue breastfeeding because they have less sleep loss [16]. In another study, women who breastfed exclusively averaged 30 minutes more nocturnal sleep than women who used formula at night. Thus new mothers should be encouraged to breastfeed exclusively since this may promote sleep during postpartum recovery [17].

Long-term effects on bone mineralization

Although lactating women lose more bone in the total body and lumbar spine than nonlactating women during the first 6months postpartum, after weaning these lactating women gained significantly more bone in the lumbar spine than did nonlactating women. Thus women do lose bone during lactation but they gain bone after weaning and lactation does not appear to be a major factor in net bone loss. In a study evaluating the association between lactation and bone size and strength from 145 women at 16–20 years after the last parturition, a longer cumulative duration of lactation was associated with larger bone size and strength. Thus breastfeeding improves bone mineralization, with decreased osteoporosis and fractures in later life [18].

Bonding with the baby

Breastfeeding appears to enhance "mothering behavior" including longer eye contact, and caressing their babies more [19]. In a randomized controlled trial on 488 infants weighing less than 2001 g, kangaroo mother care was provided in one group and traditional care was provided for another group. The investigators observed a change in the mother's perception of her child attributable to the skin-to-skin contact in the kangaroo caring position. Although this was a subjective "bonding effect" it was also recognized that in stressful situations when the infant has to remain in the hospital longer, mothers practicing kangaroo mother care felt more competent than mothers in the traditional care group. These results suggest that kangaroo mother care should be promoted actively and that mothers should be encouraged to use it as soon as possible during the intensive care period in preterm infants up to 40 weeks of gestational age [20].

References

[1] Uvnäs-Moberg K, Widström AM, Werner S, Matthiesen AS, Winberg J. Oxytocin and prolactin levels in breast-feeding women. Correlation with milk yield and duration of breast-feeding. Acta Obstet Gynecol Scand 1990;69(4):301–6.

[2] Uvnäs Moberg K, Ekström-Bergström A, Buckley S, Massarotti C, Pajalic Z, Luegmair K, et al. Maternal plasma levels of oxytocin during breastfeeding - a systematic review. PLoS One 2020;15(8):e0235806.

[3] Sobhy SI, Mohame NA. The effect of early initiation of breast feeding on the amount of vaginal blood loss during the fourth stage of labor. J Egypt Public Health Assoc 2004;79(1–2):1–12.

[4] Labbok MH. Effects of breastfeeding on the mother. Pediatr Clin North Am 2001;48(1):143–58.

[5] Vekemans M. Postpartum contraception: the lactational amenorrhea method. Eur J Contracept Reprod Health Care 1997;2(2):105–11.

[6] Trussell J. Contraceptive failure in the United States. Contraception 2011;83(5):397–404.

[7] Brown EJ, Deshmukh P, Antell K. Contraception update: oral contraception. FP Essent 2017;462:11–19.

[8] Stuebe AM, Grewen K, Meltzer-Brody S. Association between maternal mood and oxytocin response to breastfeeding. J Womens Health (Larchmt) 2013;22(4):352–61.

[9] Uvnas-Moberg K, Petersson M. Oxytocin, a mediator of anti-stress, well-being, social interaction, growth and healing. Z Psychosom Med Psychother 2005;51(1):57–80.

[10] Dewey KG. Energy and protein requirements during lactation. Annu Rev Nutr 1997;17:19–36.

[11] Peters SAE, Yang L, Guo Y, Chen Y, Bian Z, Du J, et al. Breastfeeding and the risk of maternal cardiovascular disease: a prospective study of 300000 Chinese women. J Am Heart Assoc 2017;6(6):e006081.

[12] Victora CG, Bahl R, Barros AJ, França GV, Horton S, Krasevec J, et al. Breastfeeding in the 21st century: epidemiology, mechanisms, and lifelong effect. Lancet 2016;387(10017):475–90.

[13] Liu B, Jorm L, Banks E. Parity, breastfeeding, and the subsequent risk of maternal type 2 diabetes. Diabetes Care 2010;33(6):1239–41.

[14] Jordan SJ, Na R, Johnatty SE, Wise LA, Adami HO, Brinton LA, et al. Breastfeeding and endometrial cancer risk: an analysis from the epidemiology of Endometrial Cancer Consortium. Obstet Gynecol 2017;129(6):1059–67.

[15] Li DP, Du C, Zhang ZM, Li GX, Yu ZF, Wang X, et al. Breastfeeding and ovarian cancer risk: a systematic review and meta-analysis of 40 epidemiological studies. Asian Pac J Cancer Prev 2014;15(12):4829–37.

[16] Doan T, Gardiner A, Gay CL, Lee KA. Breast-feeding increases sleep duration of new parents. J Perinat Neonatal Nurs 2007;21(3):200–6.

[17] Doan T, Gay CL, Kennedy HP, Newman J, Lee KA. Nighttime breastfeeding behavior is associated with more nocturnal sleep among first-time mothers at one month postpartum. J Clin Sleep Med 2014;10(3):313–19

[18] Wiklund PK, Xu L, Wang Q, Mikkola T, Lyytikäinen A, Völgyi E, et al. Lactation is associated with greater maternal bone size and bone strength later in life. Osteoporos Int 2012;23(7):1939–45.
[19] Britton JR, Britton HL, Gronwaldt V. Breastfeeding, sensitivity, and attachment. Pediatrics 2006;118(5):e1436–43.
[20] Tessier R, Cristo M, Velez S, Giron M, de Calume ZF, Ruiz-Palaez JG, et al. Kangaroo mother care and the bonding hypothesis. Pediatrics 1998;102(2):e17.

CHAPTER THIRTEEN

Challenges of incorporating human milk bioactive components into formulas

""If a multinational company developed a product that was a nutritionally balanced and delicious food, a wonder drug that both prevented and treated disease, cost almost nothing to produce and could be delivered in quantities controlled by the consumers' needs, the very announcement of their find would send their shares rocketing to the top of the stock market. The scientists who developed the product would win prizes and the wealth and influence of everyone involved would increase dramatically. Women have been producing such a miraculous substance, breast milk, since the beginning of human existence"."

Gabrielle Palmer

Key points

1. Enhancement of commercial formulas with bioactive components found in human milk may provide benefits in terms of immunoprotection, neurodevelopment, and metabolism.
2. Adding breast milk components to commercial formulas in an appropriate manner poses numerous scientific and regulatory challenges to the industry.

Introduction

For myriad reasons, not all infants can breastfeed. Many professional societies including the America Academy of Pediatrics, the World Health Organization, and the United States government recommend that human milk be the sole source of nutrition for infants up to 6 months of age [1–3]. Still, only about 25% of infants in the United States meet the current recommendations [4]. The reasons behind this dismal breastfeeding

The Science of Human Milk.
DOI: https://doi.org/10.1016/B978-0-323-99913-7.00014-9

rate include cultural, societal, and psychological factors. Lack of access to maternity support in the workplace likely plays a significant role [5]. There are also misconceptions about the adequacy of formula substitutes [6]. Based on the continued use of formulas in infants, optimization of the infant formulas fed to those infants who cannot be breastfed is critical and this has led to the search for bioactive ingredients from the human milk that could improve infant health and survival.

Both human milk and infant formulas provide basic nutrition. As elaborated in Chapter 3, human milk contains numerous bioactive components such as living microbes, enzymes, cells, and immunologically active molecules that are thought to provide benefits for breastfed infants. It needs to be stated that it is possible to add some of these components to formulas, but the exact dynamic nature of human milk directly provided to a baby from its own mother cannot currently be duplicated in infant formulas. Despite this, it is important to strive toward improvement in infant formulas whenever possible, and this includes the addition of beneficial bioactive ingredients like those found in human milk. In this chapter we will discuss examples of bioactive ingredients that could be added to formulas and challenges faced in attempting to improve infant formulas by their addition.

When contemplating addition of human milk bioactive components to formulas, it is important to recognize that there are various nuances that need to be considered. Macro- and micronutrient composition in formulas can meet many of the overall nutritional requirements that are provided with human milk. However, attempting to mimic breast milk as a living and dynamic fluid that includes microorganisms, bioactive enzymes, immunologically competent cells, immunoglobulins that respond to environmental stimuli, as well as numerous other factors, is near an impossible task. Even the composition of human milk varies over time which includes the compositional alterations that occur in colostrum, transitional and mature milk. Furthermore, there are changes that occur within feeding and variations that respond to environmental factors [7–9].

Developmental considerations

Milk microbiota and immunity

There is an ongoing interaction between microbiota in the milk and the infant's gastrointestinal tract and the nutrients that the infant is ingesting from the mother. Microbes in milk and the intestine synthesize large numbers of metabolites, many of which are epigenetically active, hence regulating transcriptional and protein synthetic processes. Microbes play a critical role in development of the intestinal mucosal immune system. During the first 6 months after birth, there is thought to be a critical window of development that is synchronized to the development of the immune system including memory and tolerance [10,11]. A dynamic interaction between milk microbes, nutrients, and the developing immune system exists between the mother and infant that includes a third component, the microbiome, that currently cannot be reproduced in commercial formulas [12].

When contemplating addition of human milk bioactive components to formulas, it is important to recognize that there are various nuances that need to be considered. Macro- and micronutrient composition in formulas can meet many of the overall nutritional requirements that are provided with human milk. However, attempting to mimic breast milk as a living and dynamic fluid that includes microorganisms, bioactive enzymes, immunologically competent cells, immunoglobulins that respond to environmental stimuli, as well as numerous other factors, is near an impossible task. Even the composition of human milk varies over time which includes the compositional alterations that occur in colostrum, transitional and mature milk. Furthermore, there are changes that occur within feeding and variations that respond to environmental factors [7–9].

Breastfeeding is one of the most impactful factors that shape the infant microbiome starting from 3–6 months of age (Fig. 13.1) [13,14]. Although the impact on the microbiome of human milk feeding dissipates over time, it is important to understand that exposure during a critical window of early life exists so the type of microbes with certain capabilities interact with the host and shape long-term health impacts. Even if the microbiome composition begins to become similar after approximately 6 months, the early interactions with different microbes with the host could still result in major developmental differences between individuals. Although many questions remain about the human milk microbiome, there is a high likelihood for its high relative impact during early development.

In terms of adding bioactive component to infant formulas, when it comes to the microbiome, research is needed to determine how human milk components either alone or combined with other added components affect the microbiome and how this in turn relates to metabolite production.

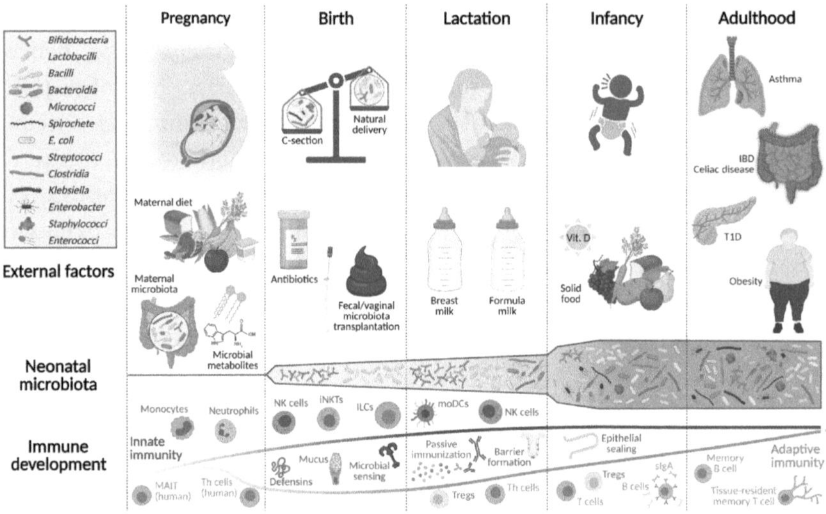

Figure 13.1 Interactions between mother, infant, the microbiome and immune responses and provision of breast milk or formula that result in long-term consequences [15]. *From Kalbermatter C, Fernandez Trigo N, Christensen S, Ganal-Vonarburg SC. Maternal microbiota, early life colonization and breast milk drive immune development in the newborn. Front Immunol 2021;12:683022.*

Simply adding certain microbes as "probiotics" to formula and expecting major benefits without an understanding of how these probiotic microbes relate to available nutritional substrates is not optimal. Furthermore, simple addition of substrates such as plant-based or even human milk oligosaccharides (HMOs) without consideration of the microbial environment lacks the level of sophistication required for understanding subsequent short- or long-term effects of these agents on the infant.

The intestinal mucosal immune system undergoes developmental maturation for the first year after birth. Exposures to antigens, microbes, and nutrients are important for development of the immune system and may play a role not only in resistance to infections but also in various allergic, atopic, or autoimmune outcomes. Some components of human milk that appear to be important for this development include immunoglobulins, the microbes, oligosaccharides, and food antigens. These may have lifelong implications for immune development and dysregulation. Clinical outcomes deemed to be important include protection against infections, oral tolerance, allergic diseases, type one diabetes, and other autoimmune diseases.

Neurodevelopment

Not only are these bioactive components important for development of the immune system, but they also have important effects on the developing brain. There are various sensitive periods during which active ingredients may have greater benefits than during other times. It is known that breastfeeding is associated with improved cognitive outcomes [16], but the precise compositional features leading to the improved cognitive outcomes remain to be discovered.

Our understanding of specific human milk bioactives and their relationship to neurobiological processes are under active investigation, much still needs to be learned. Examples of bioactives that may play a role in milk include microbes, micro RNAs, milk fat globule membranes, long-chain polyunsaturated fatty acids, HMOs, and metabolites produced by microbes including the short-chain fatty acids acetate, propionic, and butyrate (Fig. 13.2).

Breast milk composition as a reference

If breast milk's bioactive ingredient composition is to be utilized as a reference for infant formula additives, it is important to recognize

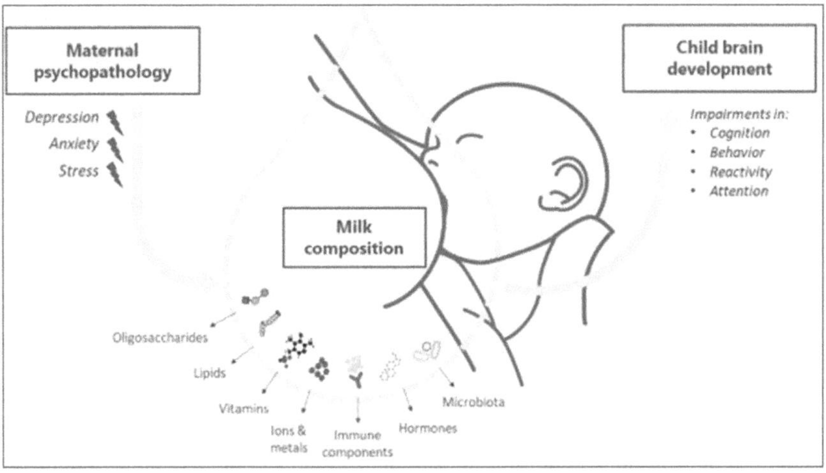

Figure 13.2 Nutritional and immunologic factors in breast milk that play a role in neurodevelopment [17]. *From Di Benedetto MG, Bottanelli C, Cattaneo A, Pariante CM, Borsini A. Nutritional and immunological factors in breast milk: a role in the intergenerational transmission from maternal psychopathology to child development. Brain Behav Immun 2020;85:57–68.*

human milk composition varies widely over the lactation period, time of day, and secondary to environmental exposures. Sex of the infant plays a role in terms of nutrient as well as bioactive ingredient requirements as well as mother's milk composition. Early nutrition has sex-specific effects and neurodevelopmental outcomes. A large same-sex twin study suggests that a mother's own milk provides sex-specific growth advantages. Formula composition does not vary with infant sex [18].

Immunoglobulins, micro-RNAs, cellular components, and numerous other factors will change over time and with various maternal and infant exposures. Precisely how important these variations are and whether a more static exposure as provided with infant formula would benefit or be detrimental to these infants remains a matter of conjecture. These outcomes are very difficult to clearly define and have become focus of recent workshops and discussions by expert panels, the National Institutes of Health as well as regulatory agencies [19,20].

Sources of bioactive ingredients added to formulas

Source of various bioactives has been a matter of consideration for the past several decades. Whether the optimal source derives from other species such as cows, plants, algae, fish, or donor human milk remains a matter of debate and is partially reliant on feasibility of the manufacturing process. Some of these factors especially the proteins can be produced in chemical synthetic processes as well as recombinant microbial synthetic techniques. Examples of bioactives that have been extensively investigated include the long-chain fatty acids derived from fish oils. These include docosahexaenoic acid (DHA), arachidonic acid (AA), and eicosatetraenoic acid. One of these, DHA, is thought to be an important component of brain development in both the fetus and the infant. It is present at 22 weeks gestation and accumulates dramatically in the brains of the infant during the first 2 years after birth [21]. Studies in nonhuman primates suggest their importance in visual acuity and other studies support their importance in cognitive function [22,23]. DHA has been added to infant formulas now for several years and research has suggested benefits, but controversy still exists [24]. Additional studies suggested that the addition of AA with DHA may be of importance [25].

Another area of interest includes the HMOs. These are found to be present in high concentration in human milk and resist digestion in the upper gastrointestinal tract. There are numerous HMOs in human milk [26]. As described in the chapter on bioactive components in human milk there are two groups of mothers that include secretors and nonsecretors [27]. Secretors have a gene that codes for a certain type of oligosaccharide called FUT2. Nonsecretors produce little to no FUT2 gene products. This provides some guidance as to types of HMOs that may incur most beneficial for a particular infant [28].

Although some formulas have already added plant-based oligosaccharides to formulas their efficacy in terms of health benefits is unclear.

Proteins and peptides

As described in the chapter on bioactives, there are numerous proteins and peptides found in human milk that appear to have major beneficial roles. These include enzymes [29] such as bile salt-stimulated lipase [30], which can assist with lipid digestion, and alkaline phosphatase, which has been implicated in playing a positive role in inactivating proinflammatory lipopolysaccharide derived from microbes [31]. Others include immunoglobulins, lactoferrin, lysozyme, and osteopontin, as discussed in a previous chapter of this book. All of these appear to play important roles in human milk, but their addition to formulas poses numerous challenges. Proteins, especially those derived from nonhuman sources have a higher potential for immunoreactivity, allergic responses, and food protein sensitivities. It is very important to (1) understand their similarity native human proteins in terms of their structure and function, (2) determine whether they are digested in the upper gastrointestinal tract or remain intact as they traverse the intestine, and (3) determine whether they are completely absorbed and have systemic actions, and their allergenic potential if manufactured as a recombinant protein or if derived from nonhuman species.

Other considerations

Addition of a single agent found in human milk to an infant formula product despite knowing their important roles in human milk may not lead to the same functionality or safety in the matrix of infant formula. Proteins may not exhibit the same quaternary folding patterns in the matrix of formula as they do in human milk. Simple addition of certain HMOs alone without appropriate microbial adjuncts may incur functions different from those in human milk. Addition of one long-chain fatty acid such as DHA without another such as in the case of DHA and AA may not be optimal in terms of overall growth of the infant. Furthermore, determination of mechanisms of action of supplemented bioactives in a formula for the infant is highly desirable but very expensive and challenging to perform.

Whenever possible the addition of bioactive components needs to critically be scrutinized for benefit and efficacy. Simply adding them because they are also found in human milk is not an adequate rationale for supplementation. Simply showing safety and not showing efficacy may be insufficient reasons for their addition. Recent workshops have discussed some of these concerns and provided guidance for families, regulatory agencies, and the industry to find the most appropriate means to proceed with adding these agents to formulas [19,20].

References

[1] Breastfeeding and the use of human milk. Pediatrics 2012;129(3):e827–41.
[2] Gomez-Pomar E, Blubaugh R. The baby friendly hospital initiative and the ten steps for successful breastfeeding. A critical review of the literature. J Perinatol 2018;38 (6):623–32.
[3] Christian P, Smith ER, Lee SE, Vargas AJ, Bremer AA, Raiten DJ. The need to study human milk as a biological system. Am J Clin Nutr 2021;113(5):1063–72.
[4] Rollins NC, Bhandari N, Hajeebhoy N, Horton S, Lutter CK, Martines JC, et al. Why invest, and what it will take to improve breastfeeding practices? Lancet 2016;387(10017):491–504.
[5] Ross E, Woszidlo A. Breastfeeding in the workplace: attitudes toward multiple roles, perceptions of support, and workplace outcomes. Breastfeed Med 2022;17 (1):38–45.
[6] Brahm P, Valdés V. The benefits of breastfeeding and associated risks of replacement with baby formulas. Rev Chil Pediatr 2017;88(1):7–14.
[7] Finn KL, Kineman BD, Czerkies LA, Carvalho RS. Human milk bioactives: future perspective. Nestle Nutr Inst Workshop Ser 2021;96:166–74.
[8] Riskin A, Almog M, Peri R, Halasz K, Srugo I, Kessel A. Changes in immunomodulatory constituents of human milk in response to active infection in the nursing infant. Pediatr Res 2012;71(2):220–5.
[9] Salminen S, Stahl B, Vinderola G, Szajewska H. Infant formula supplemented with biotics: current knowledge and future perspectives. Nutrients 2020;12(7): 1952.
[10] Gensollen T, Iyer SS, Kasper DL, Blumberg RS. How colonization by microbiota in early life shapes the immune system. Science 2016;352(6285):539–44.

[11] Cox LM, Yamanishi S, Sohn J, Alekseyenko AV, Leung JM, Cho I, et al. Altering the intestinal microbiota during a critical developmental window has lasting metabolic consequences. Cell 2014;158(4):705–21.
[12] Camacho-Morales A, Caba M, García-Juárez M, Caba-Flores MD, Viveros-Contreras R, Martínez-Valenzuela C. Breastfeeding contributes to physiological immune programming in the newborn. Front Pediatr 2021;9:744104.
[13] Ames SR, Lotoski LC, Azad MB. Comparing early life nutritional sources and human milk feeding practices: personalized and dynamic nutrition supports infant gut microbiome development and immune system maturation. Gut Microbes 2023;15(1):2190305.
[14] Andres SF, Scottoline B, Good M. Shaping infant development from the inside out: bioactive factors in human milk. Semperinatology 2023;47(1):151690.
[15] Kalbermatter C, Fernandez Trigo N, Christensen S, Ganal-Vonarburg SC. Maternal microbiota, early life colonization and breast milk drive immune development in the newborn. Front Immunol 2021;12:683022.
[16] McGowan C, Bland R. The benefits of breastfeeding on child intelligence, behavior, and executive function: a review of recent evidence. Breastfeed Med 2023;18(3):172–87.
[17] Di Benedetto MG, Bottanelli C, Cattaneo A, Pariante CM, Borsini A. Nutritional and immunological factors in breast milk: a role in the intergenerational transmission from maternal psychopathology to child development. Brain Behavior Immun 2020;85:57–68.
[18] Alur P, Ramarao S. Sex differences in preterm nutrition and growth: the evidence from human milk associated studies. J Perinatol 2022;42(8):987–92.
[19] Callahan EA, Chatila T, Deckelbaum RJ, Field CJ, Greer FR, Hernell O, et al. Assessing the safety of bioactive ingredients in infant formula that affect the immune system: recommendations from an expert panel. Am J Clin Nutr 2022;115(2):570–87.
[20] Donovan SM, Abrams SA, Azad MB, Belfort MB, Bode L, Carlson SE, et al. Summary of the joint National Institutes of Health and the Food and Drug Administration workshop titled "exploring the science surrounding the safe use of bioactive ingredients in infant formula: Considerations for an assessment framework". J Pediatr 2022;255:30–41.e1.
[21] Innis SM. Perinatal biochemistry and physiology of long-chain polyunsaturated fatty acids. J Pediatr 2003;143(4 Suppl):S1–8.
[22] Lauritzen L, Brambilla P, Mazzocchi A, Harsløf LB, Ciappolino V, Agostoni C. DHA effects in brain development and function. Nutrients. 2016;8(1):6.
[23] Cheatham CL, Colombo J, Carlson SE. N-3 fatty acids and cognitive and visual acuity development: methodologic and conceptual considerations. Am J Clin Nutr 2006;83(6 Suppl):1458S–66S.
[24] Koo WW. Efficacy and safety of docosahexaenoic acid and arachidonic acid addition to infant formulas: can one buy better vision and intelligence? J Am Coll Nutr 2003;22(2):101–7.
[25] Hadley KB, Ryan AS, Forsyth S, Gautier S, Salem Jr. N. The essentiality of arachidonic acid in infant development. Nutrients 2016;8(4):216.
[26] Bode L. Human milk oligosaccharides: every baby needs a sugar mama. Glycobiology 2012;22(9):1147–62.
[27] Soyyılmaz B, Mikš MH, Röhrig CH, Matwiejuk M, Meszaros-Matwiejuk A, Vigsnæs LK. The mean of milk: a review of human milk oligosaccharide concentrations throughout lactation. Nutrients. 2021;13(8):2737.
[28] Dinleyici M, Barbieur J, Dinleyici EC, Vandenplas Y. Functional effects of human milk oligosaccharides (HMOs). Gut Microbes 2023;15(1):2186115.

[29] Dallas DC, German JB. Enzymes in human milk. Nestle Nutr Inst Workshop Ser 2017;88:129–36.
[30] Koh J, Victor AF, Howell ML, Yeo JG, Qu Y, Selover B, et al. Bile salt-stimulated lipase activity in donor breast milk influenced by pasteurization techniques. Front Nutr 2020;7:552362.
[31] Morais J, Marques C, Faria A, Teixeira D, Barreiros-Mota I, Durão C, et al. Influence of human milk on very preterms' gut microbiota and alkaline phosphatase activity. Nutrients. 2021;13(5):1564.

CHAPTER FOURTEEN

Influence of maternal diet on breast milk composition

"A pair of substantial mammary glands has the advantage over the two hemispheres of the most learned Professor's brain, in the art of compounding a nutritious fluid for infants."

Oliver Wendall Holmes

"Mothers and babies form an inseparable biological and social unit; the health and nutrition of one group cannot be divorced from the health and nutrition of the other."

World Health Organization

Key Points

1. There is remarkable plasticity of human milk composition with relatively small changes in its composition based on normal variability in the lactating mother's diet.
2. Despite this plasticity, very large alterations in maternal diet can affect the composition of the milk and in turn, likely affect the growth and development of the infant.
3. Aroma and flavor of maternal dietary constituents *is reflected in the milk, and this may affect feeding behavior of the infant.*

Introduction

A common question about breastfeeding is whether the milk the baby receives from the mother is dependent on what the mother eats. Constituents from the mother's diet get transferred to her infant via her milk and there are considerable variations in terms of quantity and quality

The Science of Human Milk.
DOI: https://doi.org/10.1016/B978-0-323-99913-7.00015-0

of the maternal diet that can vary day to day. Despite this, lactating mothers have a remarkable capability to produce milk with a composition that is for the most part independent of her nutrient intake. However, this adaptability has its limits. There exist numerous environmental factors that can affect the lactating mother's breastfeeding capabilities and milk composition, which in turn affect her infant.

This chapter will focus on the effect of maternal nutrition on milk composition and the effects on the infant. It is important that we note some special considerations that pertain to the mother as well. For example, breastfeeding may lead to loss of maternal weight during the lactation period, but overall, lean body mass is maintained [1]. Vitamins from the mother are secreted into her milk, thus a diet deficient in vitamins may deplete the maternal stores, thus requiring an overall increase in food intake and/or dietary supplements [2]. Maternal bone minerals may decline during lactation, but remineralization usually occurs after weaning [3,4].

Factors affecting milk volume

In the first few days after birth, the milk volume is low but increases markedly after the first week postpartum [5]. When lactation is fully established, breastfeeding women produce approximately 800 mL of milk per day. Milk production increases due to various factors. These include metabolic needs of the individual infant, frequency of milk expression, and milk removal by the infant. Mothers with twins and triplets produce greater quantities of milk [6].

Some mothers have infants admitted to neonatal intensive care units and thus may not be able to provide milk to their infants directly via the breast. This may have a marked effect on the quantity and quality of milk production and often can be remedied with the advice of trained lactation consultants. Use of high-quality electric pumps, hand expression, music, breast massage, and skin-to-skin care has been found to promote the quantity of milk production [7].

Maternal dietary restriction

Does dieting (done on purpose for weight maintenance or loss) or dietary deprivation secondary to lack of available nutrition affect milk volume? Studies show that maternal dietary restriction to 1500 calories per day for short periods such as a week, as well as moderate weight loss, does not adversely affect lactation [8,9]. Similarly, mild to moderate exercise has minimal effects on milk volume and composition [10]. Extreme nutritional deprivation does, however, have a substantial effect on the volume of milk produced. Women in low-resource countries where such deprivation is common produce considerably lower volumes than in high-resource countries [11]. Factors such as maternal stress, smoking, fatigue, illness, incomplete breast emptying, and infrequent milk expression reduce milk volume [12,13].

Next, we will turn to maternal intake of ma- and micronutrients and their effects on the composition of milk.

Macronutrient nutrition

Nutrition for lactating mothers is clearly important not only to the health of the mother but also for the infant being provided her milk. Intakes of most types of food allow for elegant homeostatic mechanisms in the mother's metabolism to adjust the concentrations of various nutrients found in her milk. These concentrations frequently are in the range of those that are most conducive to the infant's health.

Macronutrients are nutrients the body needs in large amounts to function optimally. The three main macronutrients are carbohydrates, protein, and fat. Macronutrients are essential nutrients needed from the diet that provide energy and the components it needs to maintain its structure and functions. They are termed "macro" because they are required in relatively large amounts. Carbohydrates and fats are major energy sources, whereas proteins provide amino acids which are essential for building lean body mass, creating new tissue proteins in the body, and muscle repair. Breastfeeding women require approximately 500 additional kcal/day beyond that recommended for nonpregnant women [14]. Insufficient energy and/or protein provided through breast milk may result in infant growth retardation and suboptimal immunity.

Maternal diet, body composition, age, and even geographic location influence the composition of breast milk. In a systematic review and meta-analysis, it was shown that fat composition in mature milk is positively related with increments of maternal fat intake. However, protein and lactose contents in immature milk did not increase with their intakes in the lactating mothers [15]. This review suggests that avoiding high fats during lactation by the mother can prevent overnutrition of fat for the nursing infant.

Micronutrient nutrition

Micronutrients are essential dietary elements required in varying quantities, but usually in small amounts for normal growth and development. These are comprised of two major categories, vitamins and minerals. There are various micronutrients whose intake by the mother plays a critical role in the breastfed infant.

There are numeral minerals found in breast milk, but for many of these, rigorous studies are lacking. Minerals we will discuss here include calcium, zinc, and iron.

Calcium

Osteoporosis during pregnancy and lactation are common. The demands of calcium in the mother must adapt to the demand created by the fetus, the placenta, and her newborn. Women who breastfeed have a loss of calcium because of its transfer to milk. Although there is little information available on risk of fractures during pregnancy and lactation in the mother, depletion may contribute to this risk [16].

Studies show a small decrease in maternal calcium overtime during the lactation period [17]. There is little difference in calcium concentrations in the milk between term and preterm infants [18]. Certain maternal conditions such as diabetes during pregnancy do not appear to have a significant effect on milk calcium concentration [19]. One review found that in adolescent mothers, familial hypophosphatemia and hyperparathyroidism could affect calcium and phosphorus concentrations [20].

Zinc

Zinc concentrations, such as those of calcium, decline during the first days postpartum [17,21]. Similarly, certain maternal health conditions such as diabetes do not affect zinc concentrations [19]. Zinc concentrations in preterm milk are higher than in term milk and this has been hypothesized from a teleologic perspective to be an adaptation of the breast milk contents to better fit the needs of preterm infants, which need higher concentrations of zinc in smaller quantities of breast milk [22].

Iron

Iron is a part of hemoglobin and thus plays a critical role in the formation of red blood cells and the transport of oxygen to the body. It is also an important component of a variety of enzymes that play critical roles in metabolic processes. Iron needs to be provided in adequate quantities to fulfill these functions in the rapidly growing infant and if not supplied in adequate quantities results in various health issues including major neurodevelopmental delays [23]. Newborn needs for iron are met through the mobilization from hepatic reserves [17]. The estimated daily physiologic requirement for iron in infants is 0.7 mg for growth and 0.2 mg to replace basal loses [24]

Several studies have noted that term, healthy, and normal birth weight infants had sufficient iron for the formation of hemoglobin and myoglobin concomitant with growth until about 6 months of age, in fully breastfed infants [24]. Iron requirements during the first 6 months of postpartum should be provided by mother's milk intake. The iron content in mother's milk is highest in colostrum and declines during the first year of lactation. Reported median concentrations of iron in breast milk ranged from 0.04 to 1.92 mg/L (mean value—0.3 mg/L) [25,26].

Vitamin D

Very few reviews on vitamin D concentration in breast milk are available. However, available studies suggest large variations in vitamin D concentrations [22]. These variations could be partly explained by season and supplementation. Factors such as dietary intake of vitamin D, sun exposure, and skin pigmentation can have strong influences on vitamin D concentrations in milk. Technological considerations may also play a role when differences in breast milk vitamin D concentrations are measured. Milk stored in plastic containers has lower levels than those stored in glass containers, perhaps due to plastic absorption of vitamin D [27]. Vitamin D concentrations in breast milk are also significantly lower after pasteurization [27].

Micronutrients and neurodevelopmental outcomes

Studies investigating associations between breast milk levels of vitamin B6, carotenoids, or selenium and infant development suggest that pyroxidal, β-carotene, and lycopene are associated with infant neurodevelopmental outcomes. However, eligible literature is limited and there is heterogeneity in the literature that prevents conclusions on the relationship between breast milk micronutrients and infant developmental outcomes [28].

Metabolic considerations for the infant

In addition to maternal nutrition and maternal behaviors, metabolic considerations such as diabetes and obesity pre- and postnatally in the mother play considerable roles in not only breast milk composition but also in the infant's response to milk. These are frequent determinants resulting in a metabolic program that may have lifelong consequences for the infant. For example, some of these metabolic programming mechanisms relate to an increasing prevalence of neurodevelopment problems during later life in these children, as illustrated in Fig. 14.1 [29].

Metabolic disorders in the mother can include obesity, insulin resistance, and diabetes. Neurological disorders that develop in the infant include autism spectrum disorders depression and anxiety disorders. Mechanisms that cause these effects remain poorly understood; however, there are data that support a pathogenic relationship to oxidative stress, inflammation, and epigenetic mechanisms associated with early neuroplasticity leading to some of these problems. Alterations in signaling molecules such as brain-derived neurotrophic factor, gamma-aminobutyric acid, dopamine, and serotonin are thought to play important roles (Fig. 14.2) [29].

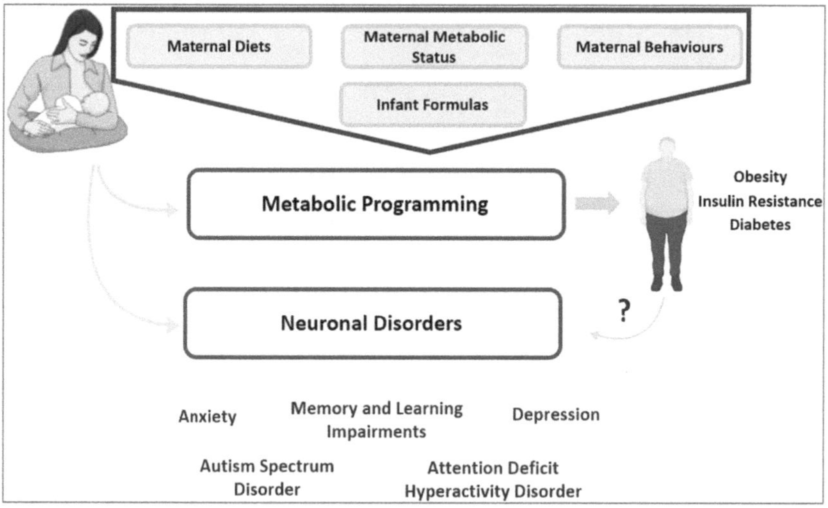

Figure 14.1 Maternal diets, behavior, and metabolic status during the lactation period that may dispose the infant for metabolic syndrome in adulthood. Various neurodevelopmental and behavioral conditions may be impacted. *From Amaro A, Baptista FI, Matafome P. Programming of future generations during breastfeeding: the intricate relation between metabolic and neurodevelopment disorders. Life Sci 2022;298:120526.*

There is clearly a critical period—a "metabolic window" that is important in the modulation of neurodevelopmental pathways [30]. Although these relationships are difficult to prove in experimental studies in humans, there are data that support these findings and thus it is crucial to be aware of the impact of maternal diet, metabolism, and their effects on the composition of milk and how this may have long-lasting effects on the offspring.

Long-chain polyunsaturated fatty acids

Long-chain polyunsaturated fatty acids (LCPUFAs) arachidonic acid (AA) and docosahexaenoic acid (DHA) levels in human milk have been given particular attention as mediators of neurodevelopmental outcomes [31,32]. Milk from well-nourished mothers had adequate levels of both AA and DHA to support the nutritional needs of infants. There is thought to be genetic variation in desaturase and elongase enzymes that are

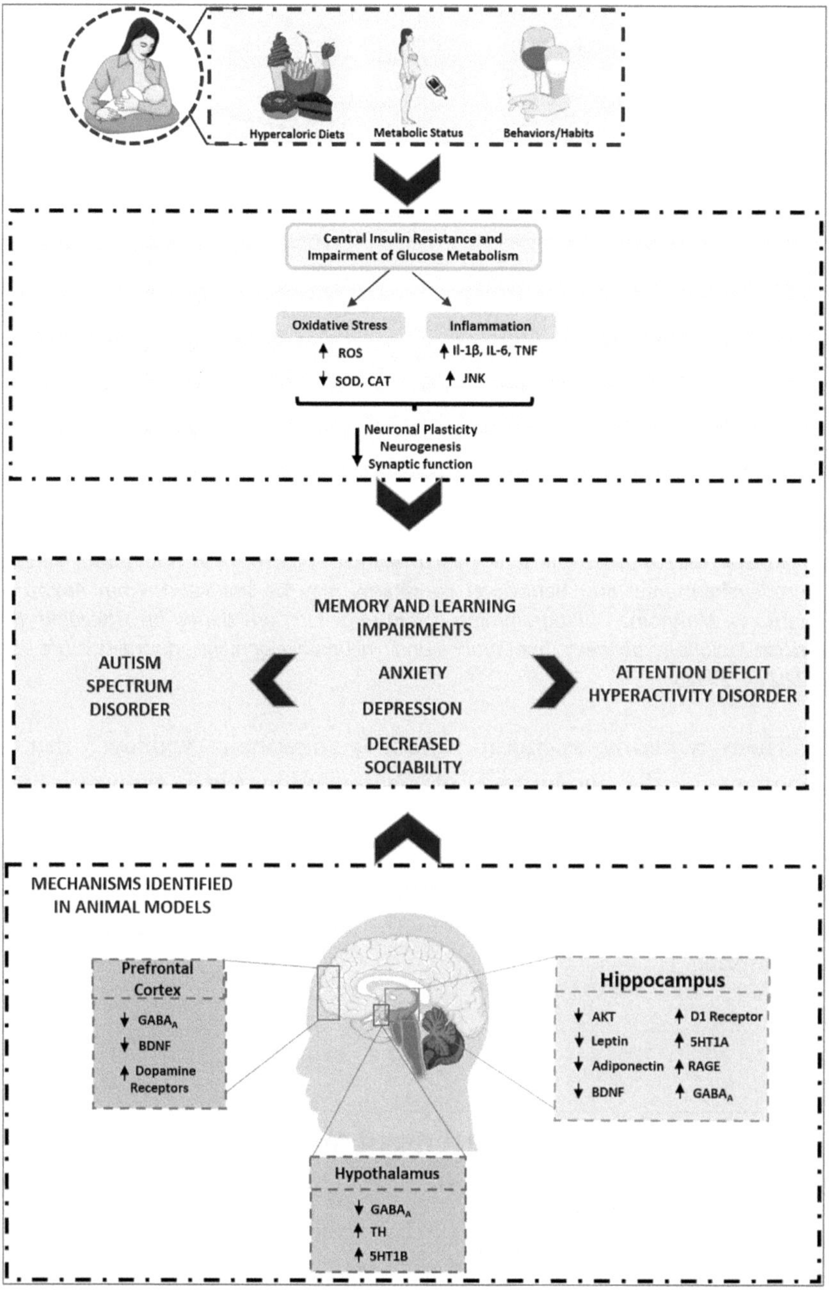

Figure 14.2 Putative mechanisms underlying alterations in neurodevelopment that related to the metabolic status and behaviors of the mother during the breastfeeding period. *From Amaro A, Baptista FI, Matafome P. Programming of future generations during breastfeeding: the intricate relation between metabolic and neurodevelopment disorders. Life Sci 2022;298:120526.*

involved in the formation of these long-chain fatty acids from their precursor 18 carbon chain length fatty acids, linoleic and linolenic acid, that has prompted addition of these LCPUFAs to formulas. The effectiveness of adding these to formulas is still being debated [31,33]. Maternal intake of these fatty acids during pregnancy and lactation has also been investigated and currently, a balanced intake of polyunsaturated fatty acids during pregnancy and lactation is recommended for the promotion of growth and development [34].

Epigenetic factors in milk

Epigenetics has been discussed in a previous chapter of this book. Here, it is added that there are interactions between maternal environment, especially mother's nutrition, the microbiome that inhabits the mother's intestine and her breast milk, and with substrates such as the oligosaccharides, vitamins such as B12, folic acid, and vitamin K that are all considered to have the potential for epigenetic consequences in the infant. It is important to note that at this time, mechanisms that occur during the lifetime of the individual are inferred largely from studies in animals, but the associations are compelling. Hence, provision of these nutrients in adequate quantities to the pregnant mother and lactating mother appears prudent.

Food aroma

Flavor and aroma of foods taken by the mother have effects on the flavor and aroma of her milk, which can in turn, affect the feeding behaviors of the infant [35]. Some foods, for instance, garlic, durian, or fenugreek, are renowned for their peculiar smells, which often affect body odor after their consumption. Maternal intake of various aromatic substances led to differences in the infant intake [36]. Vanilla led to the infant feeding longer and consuming more milk. Similar observations were made for garlic. Ethanol affected not only the infant's milk consumption but also sleep patterns were negatively affected.

Conclusion

The composition of breast milk is highly resilient to moderate changes due to the mother's diet and nutritional status. However, available data support avoidance of extremes such as stringent restrictive diets, lack of exercise or overexertion, high-stress situations, and intake of various toxins such as alcohol.

References

[1] Motil KJ, Sheng HP, Kertz BL, Montandon CM, Ellis KJ. Lean body mass of well-nourished women is preserved during lactation. Am J Clin Nutr 1998;67(2):292–300.

[2] Shinde S, Wang D, Yussuf MH, Mwanyika-Sando M, Aboud S, Fawzi WW. Micronutrient supplementation for pregnant and lactating women to improve maternal and infant nutritional status in low- and middle-income countries: protocol for a systematic review and meta-analysis. JMIR Res Protoc 2022;11(8):e40134.

[3] Grizzo FMF, Alarcão ACJ, Dell' Agnolo CM, Pedroso RB, Santos TS, Vissoci JRN, et al. How does women's bone health recover after lactation? A systematic review and meta-analysis. Osteoporos Int 2020;31(3):413–27.

[4] Kalkwarf HJ, Specker BL. Bone mineral changes during pregnancy and lactation. Endocrine 2002;17(1):49–53.

[5] Neville MC, Keller R, Seacat J, Lutes V, Neifert M, Casey C, et al. Studies in human lactation: milk volumes in lactating women during the onset of lactation and full lactation. Am J Clin Nutr 1988;48(6):1375–86.

[6] Saint L, Maggiore P, Hartmann PE. Yield and nutrient content of milk in eight women breast-feeding twins and one woman breast-feeding triplets. Br J Nutr 1986;56(1):49–58.

[7] Bendixen MM, Iapicca LC, Parker LA. Nonpharmacologic factors affecting milk production in pump-dependent mothers of critically ill infants: state of the science. Adv Neonatal Care 2023;23(1):51–63.

[8] Butte NF, Garza C, Stuff JE, Smith EO, Nichols BL. Effect of maternal diet and body composition on lactational performance. Am J Clin Nutr 1984;39(2):296–306.

[9] Dewey KG, Heinig MJ, Nommsen LA, Lonnerdal B. Maternal versus infant factors related to breast milk intake and residual milk volume: the DARLING study. Pediatrics 1991;87(6):829–37.

[10] Dewey KG, Lovelady CA, Nommsen-Rivers LA, McCrory MA, Lönnerdal B. A randomized study of the effects of aerobic exercise by lactating women on breast-milk volume and composition. N Engl J Med 1994;330(7):449–53.

[11] Brown KH, Akhtar NA, Robertson AD, Ahmed MG. Lactational capacity of marginally nourished mothers: relationships between maternal nutritional status and quantity and proximate composition of milk. Pediatrics 1986;78(5):909–19.

[12] Hopkinson JM, Schanler RJ, Garza C. Milk production by mothers of premature infants. Pediatrics 1988;81(6):815–20.

[13] Vio F, Salazar G, Infante C. Smoking during pregnancy and lactation and its effects on breast-milk volume. Am J Clin Nutr 1991;54(6):1011–16.

[14] American Academy of Pediatrics. Guidelines for perinatal care. (No Title) 2002.

[15] Xi Q, Liu W, Zeng T, Chen X, Luo T, Deng Z. Effect of different dietary patterns on macronutrient composition in human breast milk: a systematic review and meta-analysis. Nutrients. 2023;15(3):485.

[16] Lujano-Negrete AY, Rodríguez-Ruiz MC, Skinner-Taylor CM, Perez-Barbosa L, Cardenas de la Garza JA, García-Hernández PA, et al. Bone metabolism and osteoporosis during pregnancy and lactation. Arch Osteoporos 2022;17(1):36.
[17] Dror DK, Allen LH. Overview of nutrients in human milk. Adv Nutr (Bethesda, MD) 2018;9(suppl_1):278s−294ss.
[18] Gidrewicz DA, Fenton TR. A systematic review and meta-analysis of the nutrient content of preterm and term breast milk. BMC Pediatr 2014;14:216.
[19] Peila C, Gazzolo D, Bertino E, Cresi F, Coscia A. Influence of diabetes during pregnancy on human milk composition. Nutrients 2020;12(1):185.
[20] Dorea JG. Calcium and phosphorus in human milk. Nutr Res 1999;19(5):709−39.
[21] Brown KH, Engle-Stone R, Krebs NF, Peerson JM. Dietary intervention strategies to enhance zinc nutrition: promotion and support of breastfeeding for infants and young children. Food Nutr Bull 2009;30(1 Suppl):S144−71.
[22] Rios-Leyvraz M, Yao Q. Calcium, zinc, and vitamin D in breast milk: a systematic review and meta-analysis. Int Breastfeed J 2023;18(1):27.
[23] Eussen S, Alles M, Uijterschout L, Brus F, van der Horst-Graat J. Iron intake and status of children aged 6−36 months in Europe: a systematic review. Ann Nutr Metab 2015;66(2−3):80−92.
[24] Chaparro CM. Setting the stage for child health and development: prevention of iron deficiency in early infancy. J Nutr 2008;138(12):2529−33.
[25] Wasowicz W, Gromadzinska J, Szram K, Rydzynski K, Cieslak J, Pietrzak Z. Selenium, zinc, and copper concentrations in the blood and milk of lactating women. Biol Trace Elem Res 2001;79(3):221−33.
[26] Mastroeni SS, Okada IA, Rondó PH, Duran MC, Paiva AA, Neto JM. Concentrations of Fe, K, Na, Ca, P, Zn and Mg in maternal colostrum and mature milk. J Trop Pediatr 2006;52(4):272−5.
[27] Van Zoeren-Grobben D, Schrijver J, Van den Berg H, Berger HM. Human milk vitamin content after pasteurisation, storage, or tube feeding. Arch Dis Child 1987;62 (2):161−5.
[28] Lockyer F, McCann S, Moore SE. Breast milk micronutrients and infant neurodevelopmental outcomes: a systematic review. Nutrients 2021;13(11):3848.
[29] Amaro A, Baptista FI, Matafome P. Programming of future generations during breastfeeding: the intricate relation between metabolic and neurodevelopment disorders. Life Sci 2022;298:120526.
[30] Goyal MS, Iannotti LL, Raichle ME. Brain nutrition: a life span approach. Annu Rev Nutr 2018;38:381−99.
[31] Decsi T, Marosvölgyi T, Szabó É. Docosahexaenoic acid in formulas for term infants: the way from pioneer idea to mandatory dietary recommendation. Life (Basel) 2023;13(6):1326.
[32] Salem Jr. N, Van Dael P. Arachidonic acid in human milk. Nutrients 2020;12(3):626.
[33] Martinot P, Adjibade M, Taine M, Davisse-Paturet C, Lioret S, Charles MA, et al. LC-PUFA enrichment in infant formula and neurodevelopment up to age 3.5 years in the French nationwide ELFE birth cohort. Eur J Nutr 2022;61(6):2979−91.
[34] Khor GL. Implications of maternal diet on breast milk docosahexaenoic acid 22:6n-3 (DHA) and arachidonic acid 20:4n-6 (AA) contents: a narrative review. Asia Pac J Clin Nutr 2022;31(4):576−83.
[35] Debong MW, Loos HM. Diet-induced flavor changes in human milk: update and perspectives. J Agric Food Chem 2020;68(38):10275−80.
[36] Spill M, Callahan E, Johns K, Shapiro M, Spahn JM, Wong YP, et al. Influence of maternal diet on flavor transfer to amniotic fluid and breast milk and children's responses: a systematic review. Alexandria, VA: USDA Nutrition Evidence Systematic Review; 2019.

Index

Note: Page numbers followed by "*f*" and "*t*" refer to figures and tables, respectively.

C

I

K

L

M